AFFINE AND PROJECTIVE GEOMETRY

AFFINE AND PROJECTIVE GEOMETRY

M. K. Bennett

Department of Mathematics and Statistics
University of Massachusetts
Amherst, Massachusetts

A Wiley-Interscience Publication

JOHN WILEY & SONS, INC.

New York ■ Chichester ■ Brisbane ■ Toronto ■ Singapore

Library of Congress Cataloging in Publication Data:

Bennett, M. K. (Mary Katherine) 1940–
    Affine and projective geometry / M. K. Bennett.
        p. cm.
    "A Wiley-Interscience publication."
    Includes index.
    ISBN 0-471-11315-8
    1. Geometry, Affine.   2. Geometry, Projective. I. Title.
QA477.B46   1995
516′.4—dc20   94-44365

Printed in the United States of America

10 9 8 7 6 5 4 3 2 1

To Tom
who is always there for me

Contents

List of Examples

Chapter 6

Chapter 7

Chapter 8

Appendix A

Special Symbols

The following is a list of special symbols used in this book, together with the page on which each first appears.

Preface

The purpose of this text is to tie together three different approaches to affine and projective geometry: the algebraic (or analytic) approach of coordinate geometry, the axiomatic approach of synthetic geometry, and the lattice theoretic approach through the flats of a geometry. It provides, in one volume, a complete, detailed, and self-contained description of the coordinatization of (Desarguesian) affine and projective space and a thorough discussion of the lattices of flats of these spaces, ending with the Fundamental Theory of Projective Geometry, relating synthetic collineations, lattice isomorphisms, and vector space isomorphisms.

The text first aims to offer (for upper-level mathematics and mathematics education students) an opportunity to see, at a fairly early stage in their mathematical development, the basic correlation between two seemingly disparate but familiar branches of mathematics (synthetic geometry and linear algebra). The first five chapters are suitable for a one-semester course. They attempt to provide background and insight into this correlation, so important for prospective secondary teachers, and to develop an appreciation of the geometric nature of linear algebra. These chapters will also strengthen the student's understanding of abstract algebra, since they provide a different perspective on fields than is usually presented in an algebra course. They also will give the students who will not take an algebra course a modicum of familiarity with the most basic notions in abstract algebra.

The nine chapters are suitable for a two-semester course. The last four chapters concentrate on nonplanar spaces and vector spaces of dimension 3 or more. They use geometry to introduce the students to the rudiments of lattice theory, with the final chapter providing insights into the geometric nature of linear and semilinear bijections on vector spaces.

Throughout the text, geometry is used as a vehicle for introducing such topics as finite fields, some basic combinatorics, closure spaces, lattices, and algebraic systems such as division rings and those of Veblen and Wedderburn. Some of these topics may not appear in other undergraduate courses, and they may inspire some students to further investigation. There are very interesting open questions in affine and projective geometry that can easily be understood by students at this level. Discussing such questions always seems to interest students and whet their appetites for more mathematics.

There are many legitimate and interesting possibilities for a one- or two-semester course in geometry for mathematics majors. In general, geometry courses tend to be either survey courses covering several types of geometry or courses concentrating on Euclidean or one particular sort of non-Euclidean geometry. This text is of the latter variety, and focuses on affine geometry.

The main mathematical distinction between this and other single-geometry texts is the emphasis on affine rather than projective geometry. Although projective geometry is, with its duality, perhaps easier for a mathematician to study, an argument can be made that affine geometry is intuitively easier for a student. The reason for this is twofold: the Euclidean plane, already familiar, is an affine plane, and in coordinatizing affine geometry by the methods described here, one sees the actual vectors, rather than one-dimensional subspaces, in a vector space.

The use of basic algebra and linear algebra should help the students to see how one branch of mathematics can sometimes be used to prove theorems in another. The text is designed so that it can be used by students who have little or no background in algebra, as well as by students who do.

I owe thanks to several people who read earlier drafts of this text and made helpful comments. Joseph Bonin of George Washington University, Richard Greechie of Louisiana Tech University, and Robert Piziak of Baylor University used earlier drafts with their students and made welcome suggestions. I am particularly grateful to Garrett Birkhoff of Harvard University, who encouraged me in this project and who made many invaluable comments, sharing with me his wisdom and his many years of experience in mathematical writing.

For their patience and good cheer, and for the help they've given me, I would like to thank the staffs of the University of Massachusetts Physical Sciences Library and of John Wiley & Sons.

Finally, the appearance of a book gives its author the opportunity to express appreciation to those people who, while not directly involved with this project, nevertheless have had a profound influence on her mathematical life. I wish to thank Sister Eileen Flanagan who taught me the challenge of mathematics in high school, Florence D. Jacobson of Albertus Magnus College, who first made me love abstract algebra, and David J. Foulis of the University of Massachusetts, who opened for me the doors to mathematical research in lattice theory as well as affine and projective geometry. These extraordinary people have made a real difference in my life.

Amherst, Massachusetts
May 1995

AFFINE AND PROJECTIVE GEOMETRY

CHAPTER 1

Introduction

The formal study of an area of mathematics is based on *definitions*, *axioms*, and *theorems*. The basic terms needed are either *defined* or assumed to be familiar; basic facts, called *axioms*, are assumed to be true, and the theorems that can be proved using the axioms and previously proved theorems make up the mathematical content of the subject under investigation.

1.1 METHODS OF PROOF

There are two basic methods of proof: *direct proof*, and *proof by contradiction*. The following examples of mathematical proofs come from elementary number theory, first treated systematically in Euclid's *Elements*. In these examples it is assumed that the terms "integer" and "divisor" are familiar. An *even integer* is defined to be one of the form $2k$, and an *odd integer* is defined to be one of the form $2k + 1$ (k an integer). First is a simple direct proof that the product of two odd integers is odd.

■ **THEOREM 1.** If n and m are odd integers, their product $n \times m$ is also odd.

    Proof: Let $n = 2k + 1$ and $m = 2h + 1$. Then $n \times m = (2k + 1)(2h + 1) = 4kh + 2k + 2h + 1 = 2(2kh + k + h) + 1$, which is odd. ■

In a proof by contradiction, one assumes the opposite of what is to be proved and deduces an untrue statement *(or contradiction)* as in the following example:

■ **THEOREM 2.** For any integer n, if n^2 is even, then n is even.

Proof: Suppose that n is odd. Then $n = 2k + 1$ for some k; thus n^2 is the product of odd numbers and is therefore odd. This contradicts the hypothesis that n^2 is even; hence the assumption that n is odd must be false. Therefore n is even. ∎

An integer is called *prime* if it has exactly two divisors, itself and 1. Two positive integers a and b are said to be *relatively prime* if their only common divisor is 1. (In particular, when a and b are relatively prime, the fraction a/b is in lowest terms.) Take as an axiom the fact that integers a and b are relatively prime if and only if there are integers m and n with $1 = ma + nb$.

The following is a second example of a proof by contradiction:

■ **THEOREM 3.** Let p, u, and v be integers with p prime. Then, if p divides the product uv, p divides u or p divides v.

Proof: If p divides neither u nor v, then p and u are relatively prime, as are p and v. Thus $1 = mp + nu$, and $1 = m'p + n'v$. Multiplying these two equations gives $1 = (mm'p + mn'v + m'nu)p + nn'uv$. Thus p is relatively prime to uv, a contradiction. ∎

EXERCISES

1. Assume that every rational number can be written as a fraction in lowest terms. Prove that $\sqrt[2]{2}$ is irrational by assuming that $\sqrt[2]{2} = a/b$, a fraction in lowest terms.

2. Prove that $\sqrt[2]{3}$ is irrational.

*3. Prove that, for any whole number n, $\sqrt[2]{n}$ is either a whole number or is irrational.

4. By "p divides a" (written $p|a$) is meant that a can be written as mp for some integer m. Suppose that p, a, and n are positive integers with p prime and n greater than or equal to two. Prove that p divides a if and only if p divides a^n.

*5. Prove that $\sqrt[n]{p}$ is irrational if $n \geq 2$ and p is a prime number.

6. By trying some examples, discover which integers n have the property that $n|(n - 1)$.

1.2 SOME GREEK GEOMETERS

The first person whose name was actually associated to some proofs was one Thales of Miletus (supposedly a retired olive oil merchant) who lived around 600 BC. The following five theorems are now attributed to him:

1. Any diameter of a circle bisects the circle.
2. The base angles of an isosceles triangle are equal.

3. Vertical angles are equal.
4. An angle inscribed in a semicircle is a right angle.
5. If two triangles have two angles and a side of one equal respectively to the corresponding two angles and side of the other, then the triangles are congruent.

Thales' student, Pythagoras, who became even more famous than his teacher, founded a school at Crotona (in what is now southern Italy) around 540 BC. Since the members of his group (or cult) dispersed their results under the name of their founder, it is difficult to know exactly what was due to Pythagoras and what was due to his followers. The famous *Pythagorean Theorem* which bears his name states that the square of the hypoteneuse of any right triangle is equal to the sum of the squares of its legs. This theorem was known before the time of Pythagoras, but was probably first *proved* by his followers. They also maintained an interest in number lore, and hoped to explain all the mysteries (i.e., physics) of the universe in terms of the integers and fractions, but they fell upon hard times when they discovered (and tried to suppress the knowledge) that $\sqrt[2]{2}$ is irrational.

A neighboring rival school had sprouted up—the Eleatics (whose members included Zeno and Eudoxus) and these rival groups were constantly challenging and trying to best each other mathematically. Doubtless the Eleatics were delighted to draw attention to $\sqrt[2]{2}$ at every possible opportunity. The Eleatics were the precursors of the discoverers of the idea of a *limit* which is the foundation on which calculus is based. One "proof" that the area of a circle is πr^2 goes as follows: If a circle is divided into sectors (as shown in Fig. 1.1), then the area of the circle is the sum of the areas of the sectors. If the sectors are made smaller and smaller, they resemble miniscule triangles whose areas are each $\frac{1}{2}bh$. Adding all the b's gives the circumference of the circle; each h is equal to the radius of the circle. Thus

$$A = \frac{1}{2}r \times (\text{Sum of the bases})$$

$$= \frac{1}{2}r \times (\text{Circumference of circle})$$

$$= \frac{1}{2}r \times (2\pi r)$$

$$= \pi r^2.$$

The so-called proof has two things to recommend it. It seems reasonable, and it gives the right answer. All it lacks is logical rigor in passing

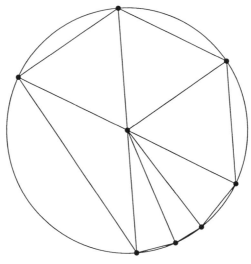

Figure 1.1.

from sectors to triangles. This was supplied centuries later when limits were discovered.

The person who contributed most to Greek geometry was not mainly an innovator at all, but an organizer and expositor. Euclid's name is probably known to more people than that of any other mathematician. His *Elements* was used as *the* geometry text for over two thousand years, and he gave his name to the "Euclidean geometry" studied in high school and "Euclidean space" studied in linear algebra to this day. Euclid was at the University of Alexandria (the first real university) around 300 BC, and undertook the enormous task of collecting and arranging in logical order all of the geometry (as well as number theory) known then. The result of his labors, Euclid's *Elements*, has been studied, copied, criticized, and generalized but certainly never ignored. Euclid started from five axioms (or self-evident properties of *magnitudes*) and five postulates (self-evident *geometrical* truths):

AXIOMS

1. Things equal to the same thing are equal to each other.
2. If equals are added to equals, their sums are equal.
3. If equals are subtracted from equals, their differences are equal.
4. Things that coincide with one another are equal to one another.
5. The whole is greater than any of its parts.

POSTULATES

1. A straight line can be drawn between any two points.
2. A line can be extended indefinitely in either direction.
3. A circle can be described with any point as center and any segment as radius.
4. All right angles are equal.
5. Through a point not on a line there exists a unique line parallel to the given line.

The fifth or *parallel* postulate (a later formulation due to John Playfair is given here) puzzled mathematicians of the eighteenth and nineteenth centuries. Until around 1830 mathematicians tried to prove it from the other axioms, and it is part of the genius of Euclid that he recognized that it had to be assumed. When this postulate is dropped or changed, different kinds of geometries result, but this was not recognized until 2000 years after Euclid's work.

Much has been written about the flaws in Euclid's logic; the essential points are these: Euclid did not have set theory for describing undefined terms such as "point" and "line," the algebra of real and rational numbers was not yet developed, and some of his postulates could have been more rigorously stated. For example, mathematicians feel that Euclid tacitly assumed the uniqueness of the line drawn in the first postulate. The visual appeal in Euclid's proofs was both a strength and a weakness. Its strength was as an aid to intuition; however, it is well known that Euclid's procedures, when applied to skillfully *mis*drawn figures, can produce *false* results, such as "every triangle is isosceles" (Kline 1972, pp. 1006–1007). Other examples may be found in (Northrup 1944, ch. 6).

This section concludes with a formal example of a geometric theorem and its direct proof based on Euclid's Axioms and Postulates, and Thales' Theorems 2 and 5 given above.

■ **THEOREM 4.** Any altitude of an equilateral triangle divides it into two congruent triangles.

Proof: Let A, B, and C be the vertices of an equilateral triangle, and let D be the endpoint of its altitude $\overline{AD}$. Then

$$\angle B = \angle C \qquad \text{(Thales' Theorem 2)},$$

$$\angle ADB = \angle ADC \qquad \text{(Euclid's Postulate 4)},$$

$$\overline{AD} = \overline{AD} \qquad \text{(Euclid's Axiom 4)},$$

$$\triangle ADB \cong \triangle ADC \qquad \text{(Thales' Theorem 5)}. \qquad ■$$

Observe that the first sentence of the proof *names* a triangle and its relevant parts. Generally, it is worthwhile to proceed in this way so that specific and rigorous statements can then be given. In the rest of the proof above, each statement was accompanied by a legitimate reason (one of Euclid's assumptions or one of Thales' theorems). The student without much experience in proving theorems is strongly encouraged to write proofs carefully, making sure that every statement is based on axioms or previously proved theorems. The best was (in fact the only way) to learn to prove theorems is by practice, using the examples given in the text as models.

EXERCISES

In the following exercises, assume Euclid's Axioms and Postulates, the theorems of Thales given above, and the Pythagorean Theorem.

1. Prove that if a line bisects an angle of a triangle and is perpendicular to the opposite side, then it bisects that side.
2. Prove that in an isosceles triangle, altitudes to the equal sides are equal.
3. Prove that the area of any equilateral triangle, each of whose sides is s units in length, is $s^2\sqrt{3}/4$.
4. Derive the formula for the area of a trapezoid with (parallel) bases b_1 and b_2 whose height is h.
5. Assuming that lines are parallel exactly when they have empty intersection, prove that if a line intersects one of two parallel lines, it intersects the other.

1.3 CARTESIAN GEOMETRY

In the seventeenth century the French philosopher and mathematician René Descartes published *La Géometrie* in which he sketched what is today called *analytic* (or *Cartesian*) geometry. The idea is quite simple. One can associate to each point of the Euclidean plane, a pair of real numbers (or *coordinates*) in such a way that every straight line is represented by a linear equation of the form $ax + by = c$, where a, b, and c are constants (a and b not both zero). Circles and conic sections are represented by quadratic equations of the form $ax^2 + bxy + cy^2 + dx + ey = f$.

Descartes began his book by asserting that "any problem in geometry can be reduced to such terms that one only needs to know the lengths of certain straight lines" to solve it. He discussed the operations of addition, subtraction, multiplication, division, and extraction of square roots, which can be performed on lengths of segments, and pointed out how easy it is to add and multiply lengths geometrically (as Euclid had done) by what is today called *geometric algebra*. Descartes' rectangular x- and y-axes intersect at an *origin*; interestingly he failed to mention the symmetry between them.

More formally, the real coordinate plane $\mathbb{R}^2$, has as its points $\{(x, y): x, y \in \mathbb{R}\}$, while its lines are the sets of points satisfying *linear equations* of the form $ax + by = c$, where a and b are not both zero. We will prove some basic results about the real coordinate plane, and show that the same results hold if the real numbers are replaced by the rational numbers, as well as by some other number systems.

■ **THEOREM 5.** In the real coordinate plane, any two distinct points are on a line.

Proof: Suppose that (x_1, y_1) and (x_2, y_2) are distinct points. We must find real numbers a, b, and c, with a and b not both zero, such that $ax + by = c$ when $(x, y) = (x_1, y_1)$ or (x_2, y_2).

The case $b = 0$: If $x_1 = x_2$, then both (x_1, y_1) and (x_2, y_2) satisfying $x = x_1$, the equation for a *vertical line*.

If $x_1 \neq x_2$, then we get the line of finite slope $\lambda = (y_2 - y_1)/(x_2 - x_1)$ and derive its equation by writing the slope using the general point (x, y) with (x_1, y_1), and setting that slope equal to λ, which gives

$$\frac{y - y_1}{x - x_1} = \frac{y_2 - y_1}{x_2 - x_1},$$

or

$$(y_2 - y_1)x + (x_1 - x_2)y = (y_2 - y_1)x_1 + (x_1 - x_2)y_1. \quad (1.1)$$

By substituting $(x, y) = (x_1, y_1)$ and $(x, y) = (x_2, y_2)$ in (1.1), it follows that both (x_1, y_1) and (x_1, y_2) satisfy (1.1).   ■

All that was needed to derive equation (1.1) were the arithmetic operations *addition*, *subtraction*, *multiplication*, and *division* (by a nonzero element) in $\mathbb{R}$. All of these operations can be restricted to the rational numbers; thus, if x_1, x_2, y_1, and y_2 were *rational* numbers, the coefficients in (1.1) would be rational as well. Thus, defining the points of the *rational coordinate plane* to be $\mathbb{Q}^2 = \{(x, y): x, y \in \mathbb{Q}\}$, where $\mathbb{Q}$ is the set of rational numbers, and its lines to be sets of points satisfying linear equations of the form $ax + by = c$, where a, b, and c are in $\mathbb{Q}$, with a and b not both zero, we have

■ **THEOREM 6.** Any two distinct points in the rational coordinate plane are contained in a line.

If two linear equations are satisfied by exactly the same points, the lines they determine are *equal* as sets. In this case their equations are *equivalent*. Two linear equations are said to be *proportional* if one is a nonzero multiple of the other; that is, if they are of the forms $ax + by = c$ and $max + mby = mc$ where m is different from zero. Since multiplication is commutative $max + mby = mc$ is the same equation as $amx + bmy = cm$ when a, b, c, and m are real (or rational) numbers.

■ **THEOREM 7.** Two lines in the rational or the real coordinate plane, whose equations are proportional, are equal lines.

Proof: If (x_1, y_1) satisfies $ax + by = c$, then $ax_1 + by_1 = c$, so $max_1 + mby_1 = mc$, and (x_1, y_1) satisfies $max + mby = mc$ for any m. Conversely, if $m \neq 0$ and $max_1 + mby_1 = mc$, then multiplying by m^{-1} gives $ax_1 + by_1 = c$. ■

■ **THEOREM 8.** In the rational or the real coordinate plane, if two lines have two points in common, they are equal.

Proof: Suppose that $(x_1, y_1) \neq (x_2, y_2)$. If both points satisfy a linear equation $ax + by = c$, then

$$ax_1 + by_1 = c \tag{1.2}$$

and

$$ax_2 + by_2 = c. \tag{1.3}$$

Subtracting gives

$$a(x_1 - x_2) - b(y_1 - y_2) = 0. \tag{1.4}$$

If $x_1 = x_2$, then $b(y_1 - y_2) = 0$, and since $y_1 \neq y_2$, $b = 0$, and the equation is of the form $ax = c$ where $a \neq 0$. But the value of c depends on the value of a, since $ax_1 = c$ by (1.2). Thus any equation satisfied by (x_1, y_1) and (x_1, y_2) is of the form $ax = ax_1$. By Theorem 7 any two such equations are proportional, so their lines are equal.

If $x_1 \neq x_2$, then for $\lambda = (y_1 - y_2)(x_1 - x_2)^{-1}$,

$$a = b\lambda. \tag{1.5}$$

Substituting in (1.2) gives

$$b\lambda x_1 + by_1 = b(\lambda + y_1) = c. \tag{1.6}$$

Thus any equation satisfied by (x_1, y_1) and (x_2, y_2) is of the form $b\lambda x + by = b(\lambda x_1 + y_1)$. By Theorem 7, any two such equations determine the same line. ■

The results in Theorems 5, 6, and 8 can be combined to give the following result:

■ **THEOREM 9.** In the rational or the real coordinate plane, any two distinct points are on one and only one line.

Definition. Two lines in a coordinate plane are said to be *parallel* if they have empty intersection. (In higher dimensions two lines are parallel if they are in the same plane and have empty intersection.)

■ **THEOREM 10.** Two lines in the rational or real coordinate plane are parallel if and only if they have equations of the form $ax + by = c$ and $ax + by = d$, where $c \neq d$.

Proof: If $c \neq d$, for no point (x_1, y_1) can it happen that $c = ax_1 + by_1 = d$, therefore no point can satisfy both $ax + by = c$ and $ax + by = d$, so the lines determined by these equations are parallel.

Conversely, the simultaneous equations

$$ax + by = c, \qquad\qquad (1.7)$$

$$a'x + b'y = c', \qquad\qquad (1.8)$$

have a common solution

$$(x, y) = \left((b'c - c'b)d^{-1}, (ac' - ca')d^{-1} \right),$$

where $d = ab' - ba'$ unless $d = 0$. In this case (assuming that $a \neq 0$) we have $b' = ba'a^{-1}$. So (1.8) becomes

$$a'x + ba'a^{-1}y = c', \qquad\qquad (1.8')$$

which for $d = c'a'^{-1}$ is proportional to

$$ax + by = d. \qquad\qquad (1.8'')$$

Equation (1.8'') is equivalent to (1.7) exactly when $c = d$. Thus (1.7) and (1.8) determine parallel lines if and only if (1.8) is equivalent to $ax + by = d$ for $c \neq d$. ■

Definition. If a line ℓ satisfies the linear equation $ax + by = c$ with $b \neq 0$, then the *slope* of ℓ is defined to be $-b^{-1}a$. If ℓ is a vertical line (i.e., if $b = 0$), ℓ is said to have *infinite slope*.

Thus any nonvertical line can be described by an equation of the form $y = mx + b$, where m is its slope and $(0, b)$ is its *y-intercept*, or point of intersection of the line with the *y*-axis.

The following result follows directly from Theorem 10:

Corollary. Two lines in the real or rational coordinate plane are parallel if and only if they have the same slope.

The parallel axiom from Euclidean geometry now becomes easy to prove in coordinate geometry. Its statement is given in Theorem 11, where we follow the convention of using capital roman letters for points and lowercase script letters for lines.

■ **THEOREM 11.** In the real or rational coordinate plane, given a point P not on a line ℓ, there is a unique line containing P and parallel to ℓ.

Proof: Suppose that $P = (p_1, p_2)$ is not on the line ℓ whose equation is $ax + by = c$. Then $c \neq ap_1 + bp_2$. Any line parallel to ℓ has an equation of the form $ax + by = d$ (where $c \neq d$). If such a line is to contain P, $d = ap_1 + bp_2$. Thus the only line containing P and parallel to ℓ is the line with equation $ax + by = ap_1 + bp_2$.

An *affine plane* is a plane in which any two points determine a unique line, and the parallel axiom holds. Theorems 9 and 11 show that the real and the rational coordinate planes are affine planes. In Chapters 2, 3, and 4 we will investigate the connections between the geometry of (synthetic) affine planes and the algebraic properties of coordinate planes.

EXERCISES

In the exercises below the capital letters P, Q, R, and so on, represent points with the corresponding lowercase coordinates, for example, $P = (p_1, p_2)$, $Q = (q_1, q_2)$. Let $\ell(P, Q)$ represent the unique line containing P and Q. Addition, subtraction, and scalar multiplication are defined as in the vector space $\mathbb{R}^2$; therefore $P + Q = (p_1 + q_1, p_2 + q_2)$ and $mP = (mp_1, mp_2)$.

1. Prove that in the real coordinate plane, if the points P, Q, and R are distinct collinear points (i.e., distinct points on the same line), then there is a nonzero real number m such that $P - Q = m(R - Q)$.

2. Suppose that P, Q, and R are noncollinear points in $\mathbb{R}^2$. Show that $\ell(P, Q)$ is parallel to $\ell(R, S)$ if and only if $P - Q$ is a nonzero multiple of $R - S$, that is, $(P - Q) = m(R - S)$ for some nonzero m.

*3. Prove that for P, Q, and R noncollinear points in the real coordinate plane, then P, Q, R, and S are the vertices of a parallelogram, with S opposite to Q, if and only if $S = P - Q + R$.

4. Prove that any linear equation with rational coefficients is equivalent to one with integer coefficients.

1.4 HILBERT'S AXIOMS

Nineteenth century geometers concentrated on *projective geometry* (to be introduced in Chapter 2). This type of geometry, in which there are no parallel lines, had its primitive beginnings when Renaissance painters were discovering the rules for painting in *perspective* to make the three-dimensional subjects of their work look realistic on two-dimensional surfaces, namely the wood panel and later the canvas. These rules are discussed, with some very interesting illustrations, in (Kline 1953, ch. 10). In the first half of the nineteenth century the problem of Euclid's parallel axiom was settled, with the result that non-Euclidean geometries in which the parallel axiom fails were of great interest to late-nineteenth-century mathematicians.

A different approach was adopted by the eminent German mathematician David Hilbert (1862–1943) who, in 1899, published his famous *Grundlagen der Geometrie*. This work was quickly translated into English and French, has gone through ten editions, and, after nearly 100 years, is still in print (Hilbert 1971)! Hilbert's aim was to make synthetic Euclidean geometry logically rigorous. To do this, he assumed 20 axioms, broken down into five groups: *incidence* (the behavior of points on lines, lines in planes, etc.), *parallelism* (the parallel axiom given above), *order* (governing "betweenness" of points on line segments), *congruence* (which introduces measurement into the geometry), and *continuity* (dealing with the topology of Euclidean geometry).

Hilbert introduced numbers as coordinates in his geometry, and described the algebraic counterparts to various geometric axioms. It is this *coordinatization* of geometries that bridges the gap between Euclid's axiomatic (or *synthetic*) and Descartes' coordinate (or *analytic*) geometry.

Hilbert's book stimulated an enormous generalization not just in geometry but in *algebra* as well. In 1900 the now common notions of abstract algebra (e.g., rings and fields) were just beginning to be developed, and Hilbert's axiomatic generalization of the Euclidean plane geometry of points and lines to the general affine plane paved the way for the accompanying algebraic generalization from the real numbers to arbitrary *fields*.

In the subsequent chapters we will follow Hilbert's lead in abstracting some parts of Euclidean geometry and applying to them the ideas of Descartes. The focus will be on affine planes, in which Euclid's incidence and parallel postulates are retained but in which there is no notion of congruence, measurement, betweenness, or the like. Thus affine geometry does not measure angles or segments, nor is it concerned with figures such as isosceles or equilateral triangles. Rather it deals with figures such as lines, triangles, and parallelograms. The aim of the next three chapters is to coordinatize affine planes, where the omission of so much of the structure of Euclidean geometry results in algebraic systems far more general than Descartes' real numbers.

1.5 FINITE COORDINATE PLANES

Besides the rational and real numbers, other algebraic systems can be used to define coordinate geometries. What is needed is a system that has a zero and a one, together with addition, subtraction, multiplication, and division (except by zero) and that satisfies the commutative, associative and distributive laws. More formally,

Definition. A *field* is a set F with two operations, addition and multiplication, satisfying the following properties for all a, b, and c in F:

$F1$. $a + b$ is in F whenever a and b are in F (closure).

$F2$. $a + b = b + a$ (commutativity).

$F3$. $a + (b + c) = (a + b) + c$ (associativity).

$F4$. There is an element 0 (called the *additive identity*) in F with $0 + a = a$ (existence of additive identity).

$F5$. For each a in F there is an element $-a$ in F with $a + (-a) = 0$ (existence of additive inverses).

$F6$. ab is in F whenever a and b are in F (closure).

$F7$. $ab = ba$ (commutativity).

$F7$. $a(bc) = (ab)c$ (associativity).

$F9$. There is an element 1 ($\neq 0$) in F (called its *multiplicative identity*) with $1a = a$ (existence of multiplicative identity).

$F10$. If $a \neq 0$, there is an element a^{-1} in F with $aa^{-1} = 1$ (existence of multiplicative inverses).

$F11$. $a(b + c) = ab + ac$; $(b + c)a = ba + ca$ (distributivity).

The most familiar examples of fields are $\mathbb{R}$ and $\mathbb{Q}$, while the simplest *finite* fields are the *integers mod p* where p is a prime number. The set $\mathbb{Z}_p$ consists of the set of *residues* $\{0, 1, ., , , . p - 1\}$, and addition and multiplication are defined as for ordinary integers and then *reduced mod p*; that is, the sum or product in $\mathbb{Z}_p$ is the *remainder* obtained when the usual sum or product is divided by p. Thus in $\mathbb{Z}_7$, $5 + 4 = 2$ (since $9 = 7 \times 1 + 2$) and in $\mathbb{Z}_5$, $4 \times 4 = 1$ (since $16 = 5 \times 3 + 1$). The smallest coordinate planes are those defined using $\mathbb{Z}_2$ and $\mathbb{Z}_3$ as given in the following examples:

□**EXAMPLE 1:** The four-point, six-line coordinate plane. The set $\mathbb{Z}_2$ consists of 0 and 1, with addition given by $0 + 0 = 1 + 1 = 0$, and $0 + 1 = 1 + 0 = 1$. Multiplication is defined by $0 \times 0 = 0 \times 1 = 1 \times 0 = 0$,

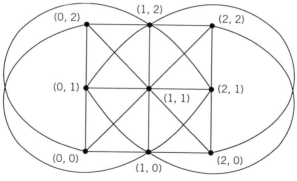

Figure 1.2.

and $1 \times 1 = 1$. Four points, $(0,0)$, $(0,1)$, $(1,0)$, and $(1,1)$, and six lines, whose equations are listed below, form the *coordinate plane over* $\mathbb{Z}_2$:

$x = 0$ contain $(0,0)$ and $(0,1)$.
$x = 1$ contains $(1,0)$ and $(1,1)$.
$y = 0$ contains $(0,0)$ and $(1,0)$.
$y = 1$ contains $(0,1)$ and $(1,1)$.
$x + y = 0$ contains $(0,0)$ and $(1,1)$,
$x + y = 1$ contains $(1,0)$ and $(0,1)$.   □

□**EXAMPLE 2: The nine-point, twelve-line coordinate plane.** $\mathbb{Z}_3 = \{0,1,2\}$ and its addition and multiplication tables are given below. The plane is represented in Fig. 1.2.

+	0	1	2		×	0	1	2
0	0	1	2		0	0	0	0
1	1	2	0		1	0	1	2
2	2	0	1		2	0	2	1

The points of the *coordinate plane over* $\mathbb{Z}_3$ are $(0,0)$, $(0,1)$, $(0,2)$, $(1,0)$, $(1,1)$, $(1,2)$, $(2,0)$, $(2,1)$, $(2,2)$. Its lines are illustrated in Fig. 1.2. Note that each equation has an equivalent form obtained by multiplying through by 2. For example, the equations $x + y = 1$ and $2x + 2y = 2$ are equivalent (proportional); therefore they determine the same line, in this case the line containing $(1,0)$, $(0,1)$ and $(2,2)$. Likewise $x + 2y = 2$ and $2x + y = 1$ are equivalent and determine the line containing $(0,1)$, $(2,0)$, and $(1,2)$.

Note that there are four sets of three parallel lines each, the "horizontal lines" whose equations are $y = 0,1,2$ are parallel, as are the "vertical lines," the "main diagonal" and the lines parallel to it, and the other "diagonal" and the two lines parallel to it. □

We will be emphasizing rational coordinate planes and finite coordinate planes that can be formed over the "modular numbers" $\mathbb{Z}_p$ (where p is prime), of which only $\mathbb{Z}_2$ and $\mathbb{Z}_3$ have been discussed here. In particular, we will define addition and multiplication based on abstract affine planes, and then will show how to represent synthetic affine planes as coordinate planes.

EXERCISES

1. Solve the following sets of simultaneous equations to find the points of intersection of the corresponding pairs of lines.
   a. $2x + 4y = 4$ and $3x + y = 0$ in the rational plane.
   b. $\pi x + 3\pi y = \sqrt[2]{7}$ and $ex + \pi y = 4$ in the real plane.
   c. $2x + y = 1$ and $x + y = 2$ in the plane over $\mathbb{Z}_3$.

2. In each case below find an equation for the line through P parallel to the line whose equation is given.
   a. $P = (3/4, 2/3)$ in the rational plane; $x + 3y = 7$.
   b. $P = (3\pi, 6)$ in the real plane; $4x + ey = 5$.
   c. $P = (1, 1)$ in the plane over $\mathbb{Z}_2$; $x + y = 1$.
   d. $P = (2, 1)$ in the plane over $\mathbb{Z}_3$; $2x + 2y = 1$.

3. Prove that the line in the real plane containing $(0, 0)$ and $(1, \sqrt[2]{2})$ contains no points but $(0, 0)$ with rational coordinates.

4. Find the points (x, y) in the plane over $\mathbb{Z}_3$ which satisfy the equation $x^2 + y^2 = 1$. Where does this "circle" intersect the line whose equation is $x = y$?

5. In general, define the unit circle in any coordinate plane as $\{(x, y): x^2 + y^2 = 1\}$. List the points on the unit circle in the plane over $\mathbb{Z}_2$ and the plane over $\mathbb{Z}_5$.

*6. Find a formula for the coordinates of the points on the unit circle in the rational plane. How does this connect with Pythagorean triples? (A *Pythagorean triple* is a set of positive integers a, b, c with $a^2 + b^2 = c^2$.)

1.6 THE THEOREMS OF PAPPUS AND DESARGUES

Two theorems of synthetic Euclidean geometry are of major importance in the coordinatization of affine planes. They will be assumed as axioms in Chapters 2 and 3.

Pappus, like Euclid a great commentator on mathematics, lived in Alexandria around 300 AD, and his *Mathematical Collection* survived long after his death. But more relevant here is the theorem that bears his name

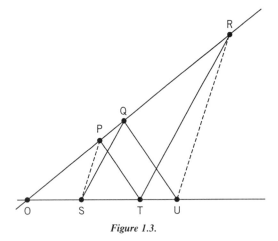

Figure 1.3.

and is illustrated in Fig. 1.3. In Fig. 1.3 in later figures, we will sometimes use dotted lines to indicate lines that are to be proved to be parallel. Assuming knowledge of the synthetic Euclidean plane, it can be proved there as follows:

■ **THEOREM 12. Pappus' Theorem.** If lines ℓ and m meet at O, with P, Q, and R in ℓ and S, T, and U in m, and if $\ell(P, T) \| \ell(Q, U)$ while $\ell(Q, S) \| \ell(R, T)$, then $\ell(P, S) \| \ell(R, U)$.

Proof: $\triangle$ PTO is similar to $\triangle$ QUO (since $\ell(Q, U)$ is parallel to the side $\ell(P, T)$ in triangle PTO). Thus $\overline{TO} / \overline{UO} = \sqrt{PO} / \overline{QO}$. Likewise $\triangle$ RTO is similar to $\triangle$ QSO, so $\overline{TO} / \overline{SO} = \overline{RO} / \overline{QO}$. Thus $\overline{UO} \cdot \overline{PO} = \overline{RO} \cdot \overline{SO}$, so that $\overline{UO} / \overline{SO} = \overline{RO} / \overline{PO}$, and since $\angle POS = \angle ROU$, $\triangle$ POS is similar to $\triangle$ ROU; therefore $\angle PSO = \angle RUO$, so $\ell(P, S) \| \ell(R, U)$.

Girard Desargues (1591–1661), an architect and friend of Descartes, proved the other Euclidean theorem we repeat here. Results from Euclidean geometry on parallelograms, as well as similar triangles, are used to prove its two parts.

■ **THEOREM 13. Desargues' Theorem (I).** Suppose that A, B, and C are distinct noncollinear points with $\ell(A, A') \| \ell(B, B') \| \ell(C, C')$, $\ell(A, B) \| \ell(A', B')$, and $\ell(A, C) \| \ell(A', C')$. Then $\ell(B, C) \| \ell(B', C')$. (See Fig. 1.4.)

Proof: Since A, C, C', and A' are the vertices of a parallelogram, the opposite sides are equal, so $\overline{AA'} = \overline{CC'}$. Similarly $\overline{AA'} = \overline{BB'}$, so $\overline{BB'} = \overline{CC'}$.

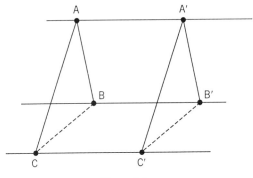

Figure 1.4.

Thus a pair of sides of B, B′, C′, C is parallel and equal, so the figure is a parallelogram, and $\ell(B, C) \| \ell(B′,′C′)$.   ■

■ **THEOREM 14. Desargues' Theorem (II).** Suppose that A, B, and C are distinct noncollinear points with $\ell(A, A′) \cap \ell(B, B′) \cap \ell(C, C′) = O$, $\ell(A, B) \| A′, B′)$, and $\ell(A, C) \| \ell(A′, C′)$. Then $\ell(B, C) \| \ell(B′, C′)$. (See Fig. 1.5.)

   Proof: Since triangles OAC and OA′C′ are similar, $\overline{OA}/\overline{OA′} = \overline{OC}/\overline{OC′}$. Since triangles OAB and OA′B′ are similar, $\overline{OA}/\overline{OA′} = \overline{OB}/\overline{OB′}$. Since $\angle BOC = \angle B′OC′$, the triangles BOC and B′OC′ are similar, so that $\ell(B, C) \| \ell(B′, C′)$.   ■

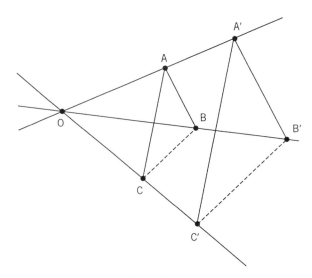

Figure 1.5.

EXERCISES

1. Give a proof of Desargues' Theorem (I) in the real coordinate plane. [*Hint:* Recall that if A, B, B', and A' are the vertices of a parallelogram, then $B' = B - A + A'$.]

2. Prove that if $(0,0)$, (a_1, a_2), and (a'_1, a'_2) are collinear in the real coordinate plane, then $(a'_1, a'_2) = m(a_1, a_2)$ for some nonzero real number m.

3. Prove that, in the real coordinate plane, if $\ell((a_1, a_2), (b_1, b_2)) \| \ell((ma_1, ma_2), (nb_1, nb_2))$ for m, $n \neq 0$, then $m = n$. [Here the notation $\ell((a_1, a_2), (b_1, b_2))$ is used for the unique line containing the points (a_1, a_2) and (b_1, b_2).

4. Prove Desargues' Theorem (II) in the real coordinate plane, assuming that O is the point $(0,0)$. [*Hint:* Use the results in Exercises 2 and 3.]

SUGGESTED READING

René Descartes, *The Geometry of René Descartes*, transl. by D. E. Smith and M.S. Latham, La Salle, IL: Open Court, 1925.

David Hilbert, *Foundations of Geometry*, 10th ed. La Salle, IL: Open Court, 1971.

Morris Kline, *Mathematical Thought from Ancient to Modern Times*, Oxford: Oxford University Press, 1972.

Morris Kline, *Mathematics in Western Culture*, Oxford: Oxford University Press, 1953.

E. P. Northrup, *Riddles in Mathematics*, New York: Van Nostrand, 1944.

CHAPTER 2

Affine Planes

It was shown in Theorems 9 and 11 of Chapter 1 that the rational and the real coordinate planes are *affine planes* in that any two points determine a unique line, and the parallel axiom holds. In Chapters 2, 3, and 4 the connections between synthetic and coordinate affine planes will be examined.

2.1 DEFINITIONS AND EXAMPLES

In addition to the planes mentioned above, the planes over $\mathbb{Z}_2$ and $\mathbb{Z}_3$ described in Chapter 1 are examples of affine planes, defined formally as follows:

Definition. An *affine plane* is an ordered pair $(\mathscr{P}, \mathscr{L})$ where $\mathscr{P}$ is a nonempty set of elements called *points* and $\mathscr{L}$ is a nonempty collection of subsets of $\mathscr{P}$ called *lines* which have the following properties:

*A*1. If P and Q are distinct points, there is a unique line ℓ such that $P \in \ell$ and $Q \in \ell$. [This line is denoted $\ell(P, Q)$.]

*A*2. If P is a point not contained in the line ℓ, there is a unique line m such that $P \in m$ and $m \cap \ell = \varnothing$. (When $\ell \cap m = \varnothing$, ℓ is said to be *parallel* to m, written $\ell \| m$.)

*A*3. There are at least two points on each line; there are at least two lines.

The second axiom defines parallel lines to be nonintersecting lines. In three-dimensional Euclidean geometry parallel lines are required to be *coplanar* as well as nonintersecting; *skew* lines are nonintersecting lines that do not lie in the same plane. Thus, assuming Axiom *A*2 ensures that the geometry is a *plane* and not a higher-dimensional space. The final axiom is included to eliminate the trivial single-point or single-line geometries.

Axioms $A1$ and $A2$ are essentially Euclid's first and fifth postulates. Furthermore any two distinct lines are either parallel or intersect in *single* point. By $A1$, if the lines ℓ and m have two distinct points in common, then $\ell = m$.

In geometry, the usual notation is opposite from that of set theory and abstract algebra; we will use capital letters to designate points and lowercase script letters for lines. When $P \in \ell$, we will sometimes say *P lies on ℓ, P is contained in ℓ,* or *ℓ passes through P.* P, Q, and R are called *collinear* if there is a line that contains all of them. In this case the line can be denoted by any of $\ell(P, Q)$, $\ell(Q, R)$, $\ell(R, Q)$, and so forth. Three or more lines are said to be *concurrent* if they intersect at a single point.

A set of axioms is *consistent* if there is a *model* or an *example* in which all the axioms hold. The existence of such a model shows that the axioms are not self-contradictory; any of the following examples of affine planes shows that the axioms are consistent.

☐ **EXAMPLE 1: The real coordinate plane.** This is the affine plane in which $\mathcal{P}$ and $\mathcal{L}$ are defined as follows:

$$\mathcal{P} = \{(x, y) : x, y \in \mathbb{R} \text{ (the real numbers)}\}$$

and

$$\ell \in \mathcal{L} \quad \text{iff} \quad \ell = \{(x, y) : ax + by = c\},$$

where $a, b, c \in \mathbb{R}$, $a^2 + b^2 \neq 0$.   ☐

☐ **EXAMPLE 2: The rational affine plane.**

$$\mathcal{P} = \mathbb{Q}^2 = \{(x, y) : x, y \in \mathbb{Q}\},$$

$$\ell \in \mathcal{L} \quad \text{iff} \quad \ell = \{(x, y) : ax + by = c\},$$

where $a, b, c \in \mathbb{Q}$, $a^2 + b^2 \neq 0$.   ☐

Theorems 9 and 11 of Chapter 1 show that $A1$ and $A2$ hold in Examples 1 and 2; $A3$ follows trivially.

☐ **EXAMPLE 3:** The smallest affine plane has four points and six lines, and is described synthetically as

$$\mathcal{P} = \{A, B, C, D\},$$

$$\mathcal{L} = \{\{AB\}, \{AC\}, \{AD\}, \{BC\}, \{BD\}, \{CD\}\}.$$

Since the lines consist of all two-element subsets of points, $A1$ (that two points lie on one and only one line) clearly holds. A check of the various possibilities verifies the parallel axiom; in fact the lines can be grouped into three sets of parallel lines: $\{AB\} \| \{CD\}$, $\{AC\} \| \{BD\}$, and $\{AD\} \| \{BC\}$. For example, A is not on $\{BD\}$, and an inspection of the lines above gives $\{AC\}$ as the only line containing A and parallel to $\{BD\}$. The third axiom for affine planes evidently holds, so $(\mathcal{P}, \mathcal{L})$ is an affine plane.   □

By setting $A = (0, 0)$, $B = (1, 0)$, $C = (0, 1)$, and $D = (1, 1)$, it follows that Example 3 can be viewed as the coordinate plane over $\mathbb{Z}_2$.

The role of each individual axiom can be clarified by exhibiting examples that satisfy two of them, but not all three. If there is an example of a geometry in which, say, $A1$ and $A2$ hold, but not $A3$, then $A3$ cannot be proved using only the first two axioms. In this case Axiom 3 is said to be *independent* of $A1$ and $A2$; a system of axioms, each of which is independent of the others, is called an *independent* system of axioms. The following three examples show that the axioms for affine planes form an independent system.

□**EXAMPLE 4: The independence of A1.** Let $\mathcal{P} = \{P, Q, R, S\}$, and let $\mathcal{L} = \{\{PS\}, \{QR\}\}$. Since there is no line containing P and Q, Axiom $A1$ fails. The two lines $\{PS\}$ and $\{QR\}$ are clearly parallel (they have empty intersection as sets), and a check of the various possible cases shows that the parallel axiom holds. For example, P is not on $\{QR\}$, and the unique line containing P and parallel to $\{QR\}$ is $\{PS\}$. The third axiom holds since there are two lines and each line contains two points. A representation of this model is given in Fig. 2.1. Note that, in the figure, even though the points P and S are connected to indicate that they are collinear, *there are no other points on that line.*   □

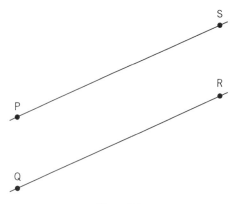

Figure 2.1

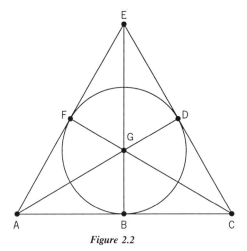

Figure 2.2

□EXAMPLE 5: The Fano plane. In 1892 Gino Fano discovered the smallest *projective plane* in which *A*1 holds, every pair of lines intersects in a point, and there are three points on each line. (See Fig. 2.2) It is the seven-point, seven-line geometry given by

$$\mathscr{P} = \{A, B, C, D, E, F, G\},$$

$$\mathscr{L} = \{\{ABC\}, \{CDE\}, \{EFA\}, \{AGD\}, \{BGE\}, \{CGF\}, \{BDF\}\}.$$

These are 21 possibilities for pairs of points, and it can be shown that any two points lie on one and only one line (e.g., C and F are on {CGF}.) The parallel axiom fails with a vengeance: There are no parallel lines, for example {AGD} ∩ {EFA} = A. Checking the 21 pairs of lines verifies that each pair of lines intersects at a point. The third axiom holds because there are three points on every line and seven lines. The Fano plane is represented in Fig. 2.2. Here it may appear that the somewhat circular line {FDB} meets {EGB} at B and somewhere between E and G. However, the sets {FDB} and {EGB} intersect only at B. This plane cannot be represented on a flat surface using straight segments for lines, since it is not part of a Euclidean geometry.   □

□EXAMPLE 6: The independence of A3. In the simple case where $\mathscr{P} = \{P, Q\}$ and $\mathscr{L} = \{\{P, Q\}\}$, *A*1 clearly holds, while the parallel axiom holds *vacuously*. In other words, the hypothesis "P is a point not on ℓ" is always false, so the parallel axiom is true. *A*3 fails because there is only one line.   □

Figure 2.3 gives two possible visualizations of the four-point, six-line affine plane described in Example 3. In the first drawing two of the pairs

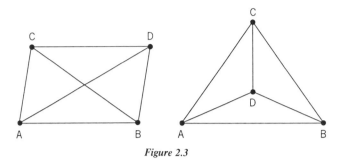

Figure 2.3

of parallel lines look parallel; the third pair *looks like* they intersect. But lines are sets of points, and two lines are parallel exactly when their intersection is empty. {BC} ∩ {AD} = ∅; thus {BC} ∥ {AD}. In the second representation, none of the pairs of parallel lines appear to intersect; however, they don't resemble Euclidean parallel lines either. As above, the difficulty is that it is impossible to make a "true-to-life" drawing of a four-point (*non-Euclidean*) affine plane on a *Euclidean* piece of paper.

Since the world we live in is *locally Euclidean* (meaning that even though the earth is not flat, it curves so gradually that in small regions it appears flat), many Euclidean theorems seem natural and unsurprising. However, theorems that hold in the Euclidean plane are not necessarily true on other surfaces. For example, the sum of the angles of a Euclidean triangle is always 180°. However, in spherical geometry, a triangle whose vertices are a pole and two points on the equator has at least two right angles! Similarly, in affine geometry, Euclidean theorems need not be true, and real world intuition can no longer be trusted. In particular, although figures can be drawn to aid in proving theorems, they are much less reliable than the figures used in Euclidean geometry, as is apparent from Figs. 2.2 and 2.3.

EXERCISES

1. Suppose that $\mathscr{P} = \{A, B, C\}$ and $\mathscr{L} = \{\{AB\}, \{AC\}, \{BC\}\}$. Show that $(\mathscr{P}, \mathscr{L})$ is not an affine plane.

2. Verify that the axioms for an affine plane hold in the following example. The points are the number-triples $(1, 0, 0)$, $(0, 1, 0)$, $(0, 0, 1)$, and $(1, 1, 1)$. The lines are the point-sets satisfying the equations $x = 0$, $y = 0$, $z = 0$, $x = y$, $x = z$, $y = z$.

3. Show that the system with points $(0, 0)$, $(1, 0)$, $(0, 1)$, and $(1, 1)$ and lines given by the equations $x = 0$, $x = 1$, $y = 0$, $y = 1$, and $x = y$ is not an affine plane.

4. Prove that the coordinate plane over $\mathbb{Z}_3$ is an affine plane.

5. Let $\mathscr{P}$ be the set of ordered pairs of integers, and let $\mathscr{L}$ be the sets of points satisfying "linear" equations of the form $ax + by = c$ (a and b integers not both zero).

   a. Find the point of intersection of the lines with equations $x - y = 0$ and $2x + 3y = 10$.

   b. Show that the lines with equations $2x + 6y = 4$ and $3x + 7y = 5$ are parallel, since they do not have a point (ordered pair of *integers*) in common.

   c. Show that $(\mathscr{P}, \mathscr{L})$ is not an affine plane.

6. Let $\mathscr{P}$ be the points in the Euclidean plane with one point removed, and let $\mathscr{L}$ be the sets of points satisfying linear equations. Show, by giving an example, that the parallel axiom fails in $(\mathscr{P}, \mathscr{L})$.

7. Prove that if a line in an affine plane is parallel to one of two intersecting lines, it intersects the other.

*8. Show that each point of any affine plane is contained in at least three lines.

2.2 SOME COMBINATORIAL RESULTS

Here we will develop some rules governing how many points a finite affine plane might have, how many lines, how many points on a line, and so forth. The plane in Example 3 has four points and six lines. Each line contains the same number (two) of points, and each set of parallel lines has the same number (again two) of lines. In Example 7, which follows, each line has three points, and each complete set of parallel lines has three lines. This is part of a more general pattern that will be studied in this section.

☐ **EXAMPLE 7. The nine-point, twelve-line affine plane:** $\mathscr{P} = \{A, B, C, D, E, F, G, H, I\}$, while the lines are the following three-point sets:

{ABC}	{AEI}	{ADG}	{CEG}
{DEF}	{BFG}	{BEH}	{BDI}
{GHI}	{CDH}	{CFI}	{AFH}

Again it should be verified that the axioms are satisfied, and this involves many cases. For example, the line containing C and I is {CFI}, and the lines are listed as four groups of three parallel lines each. {ABC} || {DEF} || {GHI}, {ADG} || {BEH} || {CFI}, and so on. This plane is represented in Fig. 2.4 (compare Fig. 2.4 with the representation of the coordinate plane over $\mathbb{Z}_3$ shown in Fig. 1.2), but once again it can't be drawn in such a way as to make its parallel lines in this plane look like *Euclidean* parallel lines.   ☐

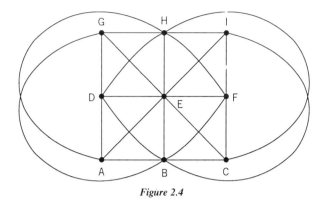

Figure 2.4

Each line in any finite affine plane has the same number of points, but more can be said. One aim here is to imitate Descartes' work in coordinatizing Euclidean geometry and bring it over to the affine setting. Consider the x and y-axes in the real coordinate plane. The x-axis consists of points of the form $(a, 0)$ where a takes on all real number values, while the y-axis consists of the points $(0, b)$. Thus the x- and y-axes correspond to the real numbers. Mapping the points on an arbitrary nonvertical line to their first coordinates gives a correspondence between the points on the line and $\mathbb{R}$; for vertical lines the second coordinates can be used.

Definition. Sets $\mathscr{S}$ and $\mathscr{T}$ are said to have the same *number of elements* (or equivalently, the same *cardinality*) when there is a one-one, onto mapping f from $\mathscr{S}$ to $\mathscr{T}$. A mapping f is *one-one* when $f(s_1) = f(s_2)$ implies that $s_1 = s_2$; f is *onto* when, given any $t \in \mathscr{T}$, there is an $s \in \mathscr{S}$ with $f(s) = t$. A one-one onto mapping is also called a *one-one correspondence*, or a *bijection* between $\mathscr{S}$ and $\mathscr{T}$. For $f : \mathscr{S} \to \mathscr{T}$ and $g : \mathscr{U} \to \mathscr{S}$, the composite $f \circ g : \mathscr{U} \to \mathscr{T}$ is given by $(f \circ g)(u) = f(g(u))$, $(u \in \mathscr{U})$, and it is well defined, since $g(u)$ is in $\mathscr{S}$.

■ **THEOREM 1**

   i. The composite of one-one mappings is one-one.

   ii. The composite of onto mappings is onto.

Proof: (i) Let f and g be one-one mappings from $\mathscr{S}$ to $\mathscr{T}$ and $\mathscr{U}$ to $\mathscr{S}$, respectively. Suppose that u and u' are distinct elements of $\mathscr{U}$. Then since g is one-one, $g(u) \neq g(u')$ in $\mathscr{S}$. Since f is one-one, $f(g(u)) \neq f(g(u'))$ in $\mathscr{T}$. Thus, by definition of a composite, this gives $f \circ g(u) \neq f \circ g(u')$, so $f \circ g$ is one-one. The proof of part ii is left as an exercise.   ■

In the Euclidean plane a one-one correspondence between the x-axis and the real numbers is given by $f((a, 0)) = a$; in fact the correspondence between the points of any line and the set of real numbers defines a one-one correspondence. Therefore Theorem 1 implies that any two lines in the real coordinate plane are in one-one correspondence, and thus have the same cardinality. As a first step in coordinatizing affine planes, the following theorem shows that any two lines in an arbitrary affine plane have the same cardinality.

■   **THEOREM 2.** Given two lines ℓ and m in an affine plane, there is a bijection between the points of ℓ and the points of m.

Proof:

CASE 1. $\ell \cap m = O \in \mathcal{P}$. Select $A \in \ell$ and $A' \in m$, both different from O. For each point $B \in \ell$ different from A and O, let k be the line containing B and parallel to $\ell(A, A')$. Now k can't be parallel to m; otherwise, there would be two lines containing A', namely m and $\ell(A, A')$ both parallel to k. Hence k intersects m at a single point, which we will call B'. Now define f from ℓ to m by $f(O) = O$, $f(A) = A'$, and $f(B) = B'$, as defined above. It is left as an exercise to check that f is a bijection.

CASE 2. $\ell \| m$. Let $O \in \ell$ and $O' \in m$. Then $\ell \cap \ell(O, O') = O$, so by Case 1 there is a one-one correspondence f from ℓ to $\ell(O, O')$; similarly there is a one-one correspondence g between $\ell(O, O')$ and m. By Theorem 1, $g \circ f$ is a one-one correspondence as well. See Fig. 2.5 for sketches of both cases.   ■

Definition. If each line of a (finite) affine plane contains exactly n points, the plane is said to have *order n*.

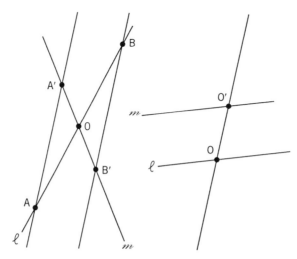

Figure 2.5

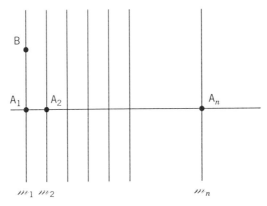

Figure 2.6

Definition. A *pencil* of parallel lines is a maximal set of mutually parallel lines, or equivalently, a set consisting of a line, together with all the lines parallel to it.

■ **THEOREM 3.** for $(\mathscr{P}, \mathscr{L})$ an affine plane of order n,

    i. $\mathscr{P}$ has exactly n^2 points,

    ii. Each point is on $n + 1$ lines,

    iii. Each pencil contains n lines,

    iv. The total number of lines is $n(n + 1)$.

    v. There are $n + 1$ pencils of parallel lines.

Proof: (i). Suppose that $\ell = \{A_1, \ldots, A_n\}$ is a line, and B is a point not on ℓ. Consider $\ell(A_1, B) = m_1$, and for $i = 2, \ldots, n$, define m_i to be the unique line through A_i parallel to m_1. Each of the lines $m_1, \ldots, m_n$ has n points, there are n of these lines, and no pair of lines has a point in common, as shown in Fig. 2.6. Therefore the number of points in $m_1 \cup \cdots \cup m_n$ is n^2. If P is any point in $\mathscr{P}$, then either $P \in m_1$ or there is a line k through P parallel to m_1. In the latter case k must intersect ℓ at one of the points $A_2, \ldots, A_n$; therefore $k = m_i$ for some i. Thus every point of $\mathscr{P}$ is in one of the lines m_i, and $\mathscr{P}$ has exactly n^2 points.

(ii) Suppose that B is an arbitrary point, and $\ell = \{A_1, \ldots, A_n\}$ is a line not containing B. The lines $\ell_i = \ell(B, A_i)$ ($i = 1, \ldots, n$) and the line ℓ_{n+1}, the line through B parallel to ℓ, are all distinct. Further, any line through B intersects ℓ (and therefore is one of the ℓ_i) or is parallel to ℓ (and therefore is ℓ_{n+1}).

(iii) Suppose that ℓ is any line, C_1 is a point on ℓ, and C_2 is a point not on ℓ. Then the points of the line $\ell(C_1, C_2)$ can be written as $C_1, \ldots, C_n$. For $i = 2, \ldots, n$, let ℓ_i be the line through C_i parallel to ℓ,

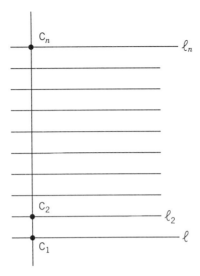

Figure 2.7

as shown in Fig. 2.7. These $n - 1$ lines are all different, since the points $C_2, \ldots, C_n$ are distinct. If k is any line parallel to ℓ, then k must intersect $\ell(C_1, C_2)$, since ℓ is the only line containing C_1 which is parallel to k. Therefore $k \cap \ell(C_1, C_2) = C_i$ for some i, so $k = \ell_i$. Consequently the lines $\ell_2, \ldots, \ell_n$ are the only lines parallel to ℓ.

(iv) Finally, suppose that B is any point, and $\ell_1, \ldots, \ell_{n+1}$ are the $n + 1$ lines containing B. Each of these lines is in a pencil (consisting of n lines), so the lines through B, together with the lines parallel to them, account for $n(n + 1)$ distinct lines. For any line m, either $B \in m$ (in which case m has been counted) or there is a line containing B and parallel to m (in which case m has also been counted). Therefore there are $n(n + 1)$ lines in $\mathcal{L}$.

(v) Since each of the $n(n + 1)$ lines is in one and only one pencil, and each pencil contains n lines, there are $n + 1$ pencils.   ∎

By Axiom 3, every finite affine plane is of order 2 or greater, and therefore contains at least three pencils of parallel lines.

EXERCISES

1. Let $(\mathcal{P}, \mathcal{L})$ be the affine plane of order three given in Example 7.
   a. List the points on the line containing G and C.
   b. List the points on the line containing B that is parallel to the line found in (a).
   c. Find an example of four distinct points, no three of which are collinear.
   d. How many pencils of parallel lines does the plane have? List the lines in the pencil containing {AEI}.

2. Prove directly from the axioms that there are at least three pencils of parallel lines in every affine plane.

3. Let $\ell \sim m$ stand for the statement "ℓ is parallel or equal to m." Prove that $\sim$ is an *equivalence relation* on the set $\mathscr{L}$ of lines of an affine plane. In other words prove each of these statements:

   a. $\ell \sim \ell$ for every line ℓ ($\sim$ is *reflexive*).

   b. If $\ell \sim m$, then $m \sim \ell$ ($\sim$ is *symmetric*).

   c. If $\ell \sim m$ and $m \sim k$, then $\ell \sim k$ ($\sim$ is *transitive*).

4. Prove that the relation $\sim$ *partitions* the set of lines of any affine plane into *equivalence classes* or pencils, that is, subsets of the form $[m] = \{\ell : \ell \sim m\}$ for a given m, in such a way that each line is in exactly one such set.

5. The proof that the composite of two one-one mappings is one-one was done by contradiction in the text. Give a direct proof of this result.

6. Give a direct proof that the composite of two onto mappings is onto.

7. Prove that if a line in an affine plane intersects one of two parallel lines, it intersects the other.

2.3 FINITE PLANES

A natural question to ask is this: For which n are there affine planes of order n? *The answer is unknown; it remains one of the major unsolved problems in affine geometry*. We have seen examples of affine planes of orders 2 and 3. It will be shown in Chapter 4 that there is an affine plane of order p^n whenever p is prime and n is a positive integer. Thus there are affine planes of order 4, 5, 7, 8, and 9. It has long been known that there is no affine plane of order 6 (such a plane would have 36 points); however, it was only shown in 1989 (by a computer search) that there is no affine plane of order 10. It is still not known whether or not there is an affine plane of order 12.

Recall that an *equivalence relation* on a set is a reflexive, symmetric, and transitive relation. Given such a relation $\pi \subset \mathscr{X} \times \mathscr{X}$, π determines a *partition* on $\mathscr{X}$, that is, a division of $\mathscr{X}$ into disjoint sets $\mathscr{A}_1, \ldots, \mathscr{A}_n, \ldots$ whose union is $\mathscr{X}$. The $\mathscr{A}_i$ are the *equivalence classes* or *cells* determined by π; elements x and y are in the same class $\mathscr{A}_i$ if and only if $x \pi y$ (meaning that the pair (x, y) is in the relation π). (More details on equivalence relations and partitions can be found, e.g., in Foulis and Munem 1988, sec. 2.3.)

Suppose that $(\mathscr{P}, \mathscr{L})$ is a finite affine plane of order n. Then $|\mathscr{P}| = n^2$, and the lines in $\mathscr{L}$ can be grouped into $n + 1$ pencils of n lines each. If $\Gamma = \{\ell_1, \ldots, \ell_n\}$ is a pencil, then each point of $\mathscr{P}$ is in exactly one line of Γ; thus Γ partitions the points of $\mathscr{P}$ into n sets (lines) with n points in each set. The equivalence relation determined by Γ, which will be denoted $\pi(\Gamma)$, is defined by $A \pi(\Gamma) B$ if $A = B$ or if $\ell(A, B) \in \Gamma$. Since each pencil determines such an equivalence relation, the plane $(\mathscr{P}, \mathscr{L})$ determines $n + 1$ equivalence relations on the set of n^2 points.

Definition. Suppose that R and S are equivalence relations on the same set $\mathscr{X}$. Then R and S are said to *commute* (or *permute*) if $R \circ S = S \circ R$ (where $x R \circ S y$ if there is an element z with $x R z$ and $z S y$). They are *complementary* if the relation $I (= \mathscr{X} \times \mathscr{X})$ is a finite composite $R \circ S \circ R \circ S \circ \cdots \circ R \circ S$, and $R \cap S = \Delta$ (where Δ is the trivial equality equivalence relation satisfying $x \Delta y$ if and only if $x = y$). R and S are *orthogonal* if they are complementary and commute (in which case $R \circ S = I$). The partitions corresponding to R and S are said to *commute*, be *complementary*, or be *orthogonal* (respectively) when R and S commute, are complementary, or are orthogonal. A set of equivalence relations (or partitions) is called an *orthogonal* set if each pair in the set is orthogonal.

■ **THEOREM 4.** If Γ and Φ are pencils of parallel lines in an affine plane, then $\pi(\Gamma)$ and $\pi(\Phi)$ are orthogonal equivalence relations on the set $\mathscr{P}$.

Proof: First, $\pi(\Gamma)$ is an equivalence relation for every pencil Γ. It was defined above to be reflexive, and it is symmetric, since $\ell(A, B) = \ell(B, A)$ for every pair of points A and B. It is transitive because for distinct points A, B, C, if $\ell(A, B)$ and $\ell(B, C)$ are in Γ, these lines are equal since only one line in Γ contains B.

If A and B are distinct points in $\mathscr{P}$ with $\ell(A, B)$ in Γ, then $A \pi(\Gamma) B$ and $B \pi(\Phi) B$, so $A \pi(\Gamma) \circ \pi(\Phi) B$. Similarly, if $\ell(A, B)$ is in Φ, $A \pi(\Gamma) A$, and $A \pi(\Phi) B$, so $A \pi(\Gamma) \circ \pi(\Phi) B$. If $\ell(A, B)$ is in neither pencil, suppose that $A \in \ell$ in pencil Γ, and $B \in m$ in Φ, and let $C = \ell \cap m$, as shown in Fig. 2.8. Then $A \pi(\Gamma) C$ and $C \pi(\Phi) B$, so again $A \pi(\Gamma) \circ \pi(\Phi) B$. Thus $\pi(\Gamma) \circ \pi(\Phi) = \mathscr{P} \times \mathscr{P}$, similarly $\pi(\Phi) \circ \pi(\Gamma) = \mathscr{P} \times \mathscr{P}$, thus the equivalence relations commute.

If A and B are distinct points, and $A \pi(\Gamma) \cap \pi(\Phi) B$, then $\ell(A, B)$ is in both Γ and Φ, a contradiction. Thus $\pi(\Gamma) \cap \pi(\Phi)$ is the equality relation, and the equivalence relations $\pi(\Gamma)$ and $\pi(\Phi)$ are complementary. ■

Corollary. If there is an affine plane $(\mathscr{P}, \mathscr{L})$ of order n, then, given a set $\mathscr{X}$ of cardinality n^2, there are partitions $\pi_1, \ldots, \pi_{n+1}$ of $\mathscr{X}$ that are pairwise orthogonal.

Proof: Set up a bijection $\phi: \mathscr{X} \to \mathscr{P}$, and define an equivalence relation on $\mathscr{X}$ for each pencil Γ by $x \equiv_\Gamma y$ if $\phi(x) \pi(\Gamma) \phi(y)$. ■

The converse to this corollary is also true. Namely, whenever a finite set has a collection of $n + 1$ orthogonal partitions, then the set has

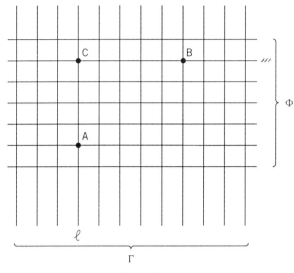

Figure 2.8

cardinality n^2, and can be considered as the set of points of an affine plane. First, however, the concept of *residues mod n* is needed.

Definition. If n is an integer greater than one, define the relation $\equiv_n$ on the set of integers $\mathbb{Z}$ by $m \equiv_n k$ if and only if n divides $m - k$. The expression $m \equiv_n k$ is read "m is congruent to k mod n."

The relation "congruence mod n" defined above is an equivalence relation. Since n divides $0 = m - m$, $\equiv_n$ is reflexive. It is symmetric, since n divides $m - k$ exactly when it divides $k - m$, and the relation is transitive since if $m \equiv_n k$ and $k \equiv_n h$, both $m - k$ and $k - h$ are divisible by n; thus $(m - k) + (k - h) = m - h$ is divisible by n.

Definition. The equivalence classes defined by the relation $\equiv_n$ are called the *congruence classes mod n.*

Each congruence class $[k]$ mod n has a unique representative from the set $\mathbb{Z}_n = \{0, 1, \ldots, n - 1\}$, called the set of *residues mod n*. The residue mod n of an arbitrary integer k is the *remainder* obtained when k is divided by n. For example, the residue mod 5 of 42 is 2, while its residue mod 9 is 6.

As was discussed in Chapter 1, where n was assumed to be prime, residues can be added and multiplied: For $i, j \in \mathbb{Z}_n$, $i +_n j$ is the residue mod n of $i + j$. Analogously $i \times_n j$ is the residue mod n of the integer ij. The operations $+_n$ and $\times_n$ are commutative and associative, since these are properties of integer addition and multiplication, and similarly multiplication distributes over addition. The residues 0 and 1 act as additive and multiplicative identifies, respectively.

■ **THEOREM 5.** The orthogonal pairs of partitions of any set correspond to the representations of this set as a Cartesian product. If the set is finite with n elements, they can be interpreted as the partitions into the rows and columns of an $m \times r$ matrix with $mr = n$.

Proof: Let π and π' be two such partitions of $\mathscr{X}$ into subsets S_i and T_j. Assign to the element $x \in \mathscr{X}$ the (unique) pair (i, j) where $x = S_i \cap T_j$—these intersections are at most singletons because $\pi \cap \pi'$ is the equality relation. Since $\pi \circ \pi' = \pi' \circ \pi = \mathscr{X}^2$, given way S_k and T_l, let $y = (k, i)$ and $z = (j, l)$. Then $y \, \pi \circ \pi' z$, so there is an x with $y \, \pi \, x \, \pi' z$. Thus the first coordinate of x is the same as that of y (i.e., k). The second coordinate of x is l, so $x = (k, l)$, and $\mathscr{X}$ is in 1-1 correspondence with the product of the subscripts assigned to the S_i and the T_j. ■

■ **THEOREM 6.** Any r-element set has *three* mutually orthogonal partitions if and only if $r = n^2$.

Proof: If $\mathscr{X}$ is partitioned by S_i, T_j, and P_l, with n, m, and k cells, respectively, then $|\mathscr{X}| = r = mn = nk = mk$; thus $m = n = k$ and $n^2 = r$.

Conversely, the points of any n^2-element set can be taken to be the pairs (i, j) of residues mod n, and the equivalence classes as the $\mathscr{S}_i$ $(i = 0, \ldots, n - 1)$ satisfying $x \equiv_n i$, the $\mathscr{T}_i$ $(i = 0, \ldots, n - 1)$ satisfying $y \equiv_n i$ and the $\mathscr{U}_i$ $(i = 0, \ldots, n - 1)$ satisfying $x + y \equiv_n i$. ■

In fact, for any k relatively prime to n, the $\mathscr{U}_i$ can be taken to be the pairs satisfying $x + ky \equiv i \pmod{n}$.

It is a simple matter to construct three pairwise orthogonal partitions of a set of n^2 numbers without reference to residues. We illustrate this with $n = 4$. Arrange the 16 numbers in a 4×4 matrix.

$$\begin{bmatrix} 1 & 2 & 3 & 4 \\ 5 & 6 & 7 & 8 \\ 9 & 10 & 11 & 12 \\ 13 & 14 & 15 & 16 \end{bmatrix}$$

Let each row be an equivalence class of π_1, and each column be an equivalence class of π_2. Construct π_3 by using the "diagonals" $\{1, 6, 11, 16\}$, $\{2, 7, 12, 13\}$, $\{3, 8, 9, 14\}$, $\{4, 5, 10, 15\}$, as the equivalence classes.

The converse to the corollary following Theorem 4 can now be proved.

■ **THEOREM 7.** Let π_i $(i = 1, \ldots, n + 1)$ be any set of $n + 1$ orthogonal partitions of a set of n^2 "points" $(n \geq 2)$. Then each partition divides the points into n equivalence classes, each consisting of n points. Moreover, if one calls the sets of points in the resulting equivalence classes "lines," the resulting points and lines form an affine plane of order n.

Proof: It follows from Theorem 6 that each line contains n points. Since the $n + 1$ lines passing through any given point must contain among them *all* of the $n^2 - 1$ points in the complement of the given point, there is a unique line through any two distinct points. The parallel axiom follows since, for C not in the π_i-equivalence class ℓ, the only line containing C and parallel to ℓ is the π_i-equivalence class containing C. This holds since if $j \neq i$, and $A \in \ell$, $A \pi_i B \pi_j$ C, so ℓ and the π_j-equivalence class containing C intersect at B. ∎

Corollary. A set of n^2 points can have at most $n + 1$ mutually orthogonal partitions.

Proof: Any π_{n+2}, added to the list above, would give *two* lines through some pair of points, so that $\pi_{n+2} \cap \pi_i$ would be different from equality for some π_i. ∎

EXERCISES

1. Show that the rational numbers can be defined as the equivalence classes determined by the relation $\approx$ on $\mathbb{Z} \times \mathbb{Z}$ where $(a, b) \approx (c, d)$ if and only if $ad = bc$.

2. a. Prove that for integers m, m', k, and k', if $m \equiv_n m'$ and $k \equiv_n k'$, then $m + m' \equiv_n k + k'$ and $mm' \equiv_n kk'$.

   b. Show that $\mathbb{Z}_n$ is closed under $+_n$, that $+_n$ is commutative and associative, that 0 acts as an identity for $+_n$, and that each element of $\mathbb{Z}_n$ has an inverse relative to $+_n$.

   c. Show that $\times_n$ is commutative and associative, that 1 is an identity for $\times_n$, and that $\times_n$ distributes over $+_n$.

   d. Show that m in $\mathbb{Z}_n$ has a multiplicative inverse in $\mathbb{Z}_n$ if and only if m and n are relatively prime.

3. Suppose that $\mathcal{Y}$ is a finite set and that π and π' are orthogonal equivalence relations on $\mathcal{Y}$. If $f : \mathcal{X} \to \mathcal{Y}$ is a bijection from a set $\mathcal{X}$ to $\mathcal{Y}$, show that σ and σ' are orthogonal equivalence relations on $\mathcal{X}$ where $x_1 \sigma x_2$ if and only if $f(x_1) \pi f(x_2)$ and $x_1 \sigma' x_2$ if and only if $f(x_1) \pi' f(x_2)$.

4. a. Show that π_1, π_2, and π_3 (as defined above) are orthogonal partitions of $\{1, 2, \ldots, 16\}$.

   b. Show that, after arranging the numbers $1, \ldots, 16$ in a 4×4 matrix, the rows, columns, and *two* sets of diagonals do not give four orthogonal partitions.

   c. Show that there is no partition of $\{1, \ldots, 16\}$ that is orthogonal to π_1, π_2, and π_3.

   d. Find another set of three orthogonal partitions of $\{1, \ldots, 16\}$.

   e. Find three orthogonal partitions of the set of the first sixteen letters $\{A, \ldots, P\}$ of the alphabet.

5. Find three mutually orthogonal partitions of the 25-element set $\mathbb{Z}_5 \times \mathbb{Z}_5$. Find a second set in which at least one of the partitions is different from all of the first three. Do the same for $\mathbb{Z}_7 \times \mathbb{Z}_7$.

6. Find three orthogonal partitions of the set of 25 letters $\{A, B, C, \dots, X, Y\}$.

2.4 ORTHOGONAL LATIN SQUARES

Toward the end of his life, the great Swiss mathematician Leonhard Euler (1707–83) was interested in the following question. Can you take 36 soliders, one officer of each of six different ranks from each of six different regiments, and arrange them in six rows and six columns so that each row contains a soldier from each rank and one from each regiment, and each column contains a soldier from each rank and one from each regiment? Euler conjectured in 1782 that this cannot be done, but he could not prove it. However, he showed that such an arrangement can be made for 9 soldiers of three ranks and three regiments; also for 16 soldiers from four ranks and four regiments, and for 25 soldiers from five ranks and five regiments. Suppose that the ranks are private, sergeant, and general, while the regiments are companies A, B, and C. The 9 soldiers are a private from each company, a sergeant from each company, and a general from each company. They can be arranged as

Company A private	Company B sergeant	Company C general
Company B general	Company C private	Company A sergeant
Company C sergeant	Company A general	Company B private

Breaking the arrangement apart, we get two 3×3 squares, one for the ranks and one for the regiments:

A	B	C		Private	Sergeant	General
B	C	A		General	Private	Sergeant
C	A	B		Sergeant	General	Private

Since what the actual ranks and regiments are called is irrelevant, we designate "Company A" and "private" by 0, "Company B" and "sergeant" by 1, and "Company C" and "general" by 2. The squares now become

0	1	2		0	1	2
1	2	0		2	0	1
2	0	1		1	2	0

Each of the residues in $\mathbb{Z}_3$ appears in each square in one row and one column. When the squares are put together, each of the pairs from

$\mathbb{Z}_3 \times \mathbb{Z}_3$ appears exactly once:

$$
\begin{array}{ccc}
(0,0) & (1,1) & (2,2) \\
(1,2) & (2,0) & (0,1) \\
(2,1) & (0,2) & (1,0)
\end{array}
$$

These concepts can be formalized as follows:

Definition.

1. A *Latin square of order n* is an $n \times n$ matrix whose entries are in the set of residues $\mathbb{Z}_n = \{0, \ldots, n-1\}$ and such that each residue occurs exactly once in each row and once in each column.
2. Latin squares $A = [a_{ij}]$ and $B = [b_{ij}]$ are *orthogonal* if, for each (k, l) in $\mathbb{Z}_n \times \mathbb{Z}_n$, there is a (necessarily unique) pair (i, j) such that $(a_{ij}, b_{ij}) = (k, l)$.
3. A collection of Latin squares is *mutually orthogonal* if every pair of them is orthogonal.

Euler was able to show that there is no pair of orthogonal Latin squares of order 2 and that there is a pair for each of the orders 3, 4, and 5. He conjectured that there is no pair of any order congruent to 2, mod 4 (6, 10, 14, etc.). However Euler's conjecture holds only for the orders 2 and 6. In 1900 Gaston Tarry (by a brute force argument) showed that there is no pair of orthogonal Latin squares of order 6. In (Betten 1984) there is a short and elegant proof of this fact. Orthogonal Latin squares of order 10 and order 22 were constructed in 1959 by Bose, Shrikhande, and Parker, and it is now known that there is a pair of orthogonal Latin squares of order n for every n except 2 and 6. More about Latin squares can be found in any book on combinatorics; see, for example, Bogart (1983) and, at a more advanced level, Hall (1986) and Dénes and Keedwell (1974).

We give here an example of a pair of orthogonal Latin squares of order 5, which is a special case of a more general construction:

$$
A = \begin{bmatrix}
0 & 1 & 2 & 3 & 4 \\
1 & 2 & 3 & 4 & 0 \\
2 & 3 & 4 & 0 & 1 \\
3 & 4 & 0 & 1 & 2 \\
4 & 0 & 1 & 2 & 3
\end{bmatrix}, \quad
B = \begin{bmatrix}
4 & 0 & 1 & 2 & 3 \\
3 & 4 & 0 & 1 & 2 \\
2 & 3 & 4 & 0 & 1 \\
1 & 2 & 3 & 4 & 0 \\
0 & 1 & 2 & 3 & 4
\end{bmatrix}.
$$

■ **THEOREM 8.** For any odd number $n = 2k + 1$, a pair of Latin squares of order n can be defined as follows: Let $A = [a_{ij}]$ be the $n \times n$ matrix defined by $a_{ij} = i +_n j$, and $B = [b_{ij}]$ be the $n \times n$ matrix whose given by $b_{ij} = i -_n j$. Then A and B are orthogonal Latin squares.

Proof: First we show that A is a Latin square. If, for some fixed i and distinct j and k, $a_{ij} = a_{ik}$, then $i + j \equiv_n i + k$. Thus n divides $(i + j) - (i + k) = j - k$, so $j \equiv_n k$. But since j and k are in $\mathbb{Z}_n$, $j = k$, and no two entries of a row of A are the same. Similarly no two entries of a column of A are equal; thus each row and each column of A contains each of the numbers $0, \ldots, (n - 1)$ exactly once.

If $i + n - j \equiv_n i + n - k$, then n divides $(i + n - j) - (i + n - k) = k - j$, so $j \equiv_n k$, and since both are in $\mathbb{Z}_n$, $j = k$. Thus no row of B has two equal entries. A similar argument for the columns of B completes the proof.

To show that A and B are orthogonal, it is enough to show that no pairs (a_{ij}, b_{ij}) and (a_{hk}, b_{hk}) are equal. If these pairs were equal, then $i +_n j = h +_n k$, and $i -_n j = n -_n k$. Adding these congruences gives $2i \equiv_n 2h$. Since n is odd, 2 and n are relatively prime. Thus 2 has a multiplicative inverse in $\mathbb{Z}_n$ so that $i \equiv_n h$, and therefore $i = h$. If $i +_n j = i +_n k$, then adding $n - i$ to both sides given $j = k$. Therefore A and B are orthogonal Latin squares. ∎

Note that A and B can be written without considering addition mod n. For A, write the first row and the first column as 0 through $n - 1$ in ascending order. Fill in the remaining rows by continuing the ascending order, cycling back to 0 when n is reached. Thus the third row is

$$3, 4, \ldots, n - 1, 0, 1, 2.$$

The rows of B are the same as those of A but are written in reverse order (thus the last row of A is the first row of B, the second-last row of A is the second row of B, etc.).

Once a pair of orthogonal Latin squares has been constructed, other pairs can be formed (meaning other pairs orthogonal to each other but not orthogonal to the original pair). For example, if A and B are orthogonal Latin squares, and J is the $n \times n$ matrix with all entries being one's, then $A + J$ and $B + J$ are orthogonal Latin squares (addition done mod n). In fact, if K is a constant matrix, each of whose entries is k, then $A + J$ and $B + K$ are orthogonal Latin squares.

The situation concerning three mutually orthogonal Latin squares is much more difficult. For example, it is not known whether there exist three mutually orthogonal Latin squares of order 10. The connection between affine planes and Latin squares will be discussed in the next section.

EXERCISES

1. Show that any permutation of the rows of a Latin square results in another Latin square. Must the two squares be orthogonal? Can they be orthogonal?

2. Show that the construction described in Theorem 8 does not work for $n = 4$.

3. Show that there is no Latin square orthogonal to

$$
\begin{array}{cccc}
0 & 1 & 2 & 3 \\
1 & 2 & 3 & 0 \\
2 & 3 & 0 & 1 \\
3 & 0 & 1 & 2
\end{array}
$$

4. Suppose that A and B are orthogonal Latin squares of order n and that J and K are constant $n \times n$ matrices with entries j and k, respectively. Prove that $A + J$ and $B + K$ are orthogonal Latin squares (again the addition should be done mod n).

5. Suppose that A and B are orthogonal Latin squares of order n, and that k in $\mathbb{Z}_n$ is relatively prime to n. Show that A and kB are orthogonal Latin squares, where the ijth entry in kB is $k \times_n b_{ij}$.

6. Let n be an odd natural number. Show that $A = [a_{ij}]$ and $B = [b_{ij}]$ are orthogonal Latin squares of order n where

$$ a_{ij} \equiv_n i + j - 1 $$

and

$$ b_{ij} \equiv_n 2i + j - 1. $$

7. Show that there are only two Latin squares of order 2, and that they are *not* orthogonal.

8. Give an example of a pair of 3×3 orthogonal Latin squares. Show that there is no third square orthogonal to both of them.

*9. Construct three mutually orthogonal Latin squares of order 4. Construct three of order 5.

2.5 AFFINE PLANES AND LATIN SQUARES

We will denote by $H(n)$ the maximal number of mutually orthogonal Latin squares of order n. In this language it was already seen that $H(2) = H(6) = 1$ and that $H(n) \geq 2$ otherwise. Results on affine planes in Chapter 4 will imply that whenever p is a prime number, $H(p^k) = p^k - 1$.

The connection between pencils of parallel lines and Latin squares is summarized in the following theorem:

■ **THEOREM 9.** For $n \geq 2$, there is an affine plane of order n if and only if $H(n) \geq n - 1$.

Proof: Given an affine plane of order n, select two of its $(n + 1)$ pencils of parallel lines, and denote them by Γ_1 and Γ_2. Denote the lines of Γ_1 by $m_0, \ldots, m_{n-1}$, and the lines of Γ_2 by $k_0, \ldots, k_{n-1}$. Then each point P in the plane can be labeled uniquely by the pair of residues (i, j) where $P = m_i \cap k_j$. For each of the remaining $n - 1$ pencils Φ, number its lines by the residues $0, \ldots, n - 1$. Form the Latin square associated with Φ by defining its (i, j)th entry to be the residue k where the point $P = (i, j)$ is in the kth line of Φ. The $n - 1$ Latin squares so obtained are mutually orthogonal.

Conversely, given a family $A_1, \ldots, A_{n-1}$ of mutually orthogonal Latin squares of order n, one can define an affine plane whose points are $\{(i, j): 0 \le i, j \le n - 1\}$ as follows. The $2n$ sets given by $x = i$ and $y = j$ are defined to be lines. The remaining $n(n - 1)$ lines are given by the sets $\{(i, j): A_k(i, j) = s\}$ for fixed A_k and residue s (where $A_k(i, j)$ denotes the entry in the ith row and jth column of A_k).

For $j \ne j'$ in $\mathbb{Z}_n$, the points (i, j) and (i, j') are on the line $x = i$. If they are also on a line determined by the Latin square A_k, then $A_k(i, j) = A_k(i, j')$ a contradiction since no row of A_k contains the same element twice.

Given two points (i, j) and (i', j'), with $i \ne i'$ and $j \ne j'$, if they are on lines determined by A_k and A_h, then $A_k(i, j) = A_k(i', j') = s$ in $\mathbb{Z}_n$ and $A_h(i, j) = A_h(i', j') = t$ in $\mathbb{Z}_n$. Thus the pair (s, t) appears twice when the squares A_k and A_h are put together, a contradiction of the orthogonality of A_k and A_h. Thus each pair (i, j) and (i', j') is in at most one line.

There are n lines determined by each A_k, and $2n$ lines of the form $x = i$ and $y = j$, so the total number of lines is $n(n + 1)$, each line containing n points. Since no pair of points is on more than one line, the total number of *collinear pairs* is given by

$$n(n + 1)\begin{bmatrix} n \\ 2 \end{bmatrix} = \frac{n(n + 1)n(n - 1)}{2} = \frac{n^2(n^2 - 1)}{2} = \begin{bmatrix} n^2 \\ 2 \end{bmatrix}.$$

Thus each of the n^2 pairs of points is a collinear pair; therefore each pair of points is on one and only one line.

The parallel axiom follows since the lines $x = i$ are mutually parallel, the lines $y = j$ are mutually parallel, and for each A_k, the n lines it determines are mutually parallel. Since n is two or greater, each line contains at least two points, and there are at least two (in fact, at least three) lines, so we have an affine plane. ∎

It is known that $H(n)$ is always less than or equal to $n - 1$; thus $H(n) = n - 1$ if and only if there is an affine plane of order n.

The construction of three orthogonal Latin squares can be illustrated by an affine plane of order 4, given in Example 8.

☐**EXAMPLE 8. The affine plane of order 4:** $\mathcal{P} = \{A, \ldots, P\}$, and the lines in $\mathcal{L}$ are grouped into pencils as shown below. Note that three of the pencils, Γ_1, Γ_2, and Φ are named.

$$
\left.\begin{array}{l}
m_0 = \{\text{OIDF}\} \\
m_1 = \{\text{ECPA}\} \\
m_2 = \{\text{KMJB}\} \\
m_3 = \{\text{GHLN}\}
\end{array}\right\}\Gamma_1
\qquad
\left.\begin{array}{l}
\ell_0 = \{\text{HBCD}\} \\
\ell_1 = \{\text{EFGH}\} \\
\ell_2 = \{\text{IAKL}\} \\
\ell_3 = \{\text{MNOP}\}
\end{array}\right\}\Phi
\qquad
\begin{array}{l}
\{\text{HFKP}\} \\
\{\text{BEOL}\} \\
\{\text{JNCI}\} \\
\{\text{AGDM}\}
\end{array}
$$

$$
\left.\begin{array}{l}
k_0 = \{\text{OAJH}\} \\
k_1 = \{\text{GIPB}\} \\
k_2 = \{\text{FCML}\} \\
k_3 = \{\text{EDKN}\}
\end{array}\right\}\Gamma_2
\qquad
\begin{array}{l}
\{\text{HEIM}\} \\
\{\text{BFAN}\} \\
\{\text{CGKO}\} \\
\{\text{DJLP}\}
\end{array}
\qquad \square
$$

The point E is in m_1 and in k_3; thus it is labeled $(1,3)$. A list of the labeling of the points as pairs of the residues $(0,1,2,3)$ follows (the parentheses are omitted for convenience):

A = 10	E = 13	I = 01	M = 22
B = 21	F = 02	J = 20	N = 33
C = 12	G = 31	K = 23	O = 00
D = 03	H = 30	L = 32	P = 11

To form a Latin square from pencil Φ, consider, for example, the point E labeled 13. Point E will determine the $(1,3)$ entry in the 4×4 matrix, namely the entry in the second row (row 1), last column (column 3). Since E is in line ℓ_1, the entry in the second row, last column is 1. The entire Latin square is listed below:

$$
\begin{array}{cccc}
3 & 2 & 1 & 0 \\
2 & 3 & 0 & 1 \\
1 & 0 & 3 & 2 \\
0 & 1 & 2 & 3
\end{array}
$$

To illustrate the reverse construction, two orthogonal Latin squares of order 3 are given below, together with a 3×3 array of the letters $A, \ldots, I$:

$$
\begin{array}{ccc}
\text{ABC} & 012 & 012 \\
\text{DEF} & 120 & 201 \\
\text{GHI} & 201 & 120
\end{array}
$$

Suppose that the points of the plane are $A = (0,0)$, $B = (0,1)$, $C = (0,2)$, $D = (1,0)$, $E = (1,1)$, $F = (1,2)$, $G = (2,0)$, $H = (2,1)$, and $I = (2,2)$. The lines given by $x \equiv i$ are $\{\text{ABC}\}$, $\{\text{DEF}\}$, $\{\text{GHI}\}$. Those given by

01234	56789	01923	84657
34012	79865	67895	23104
43120	97658	93746	58210
12407	85396	38254	79061
20375	68941	14507	36982
57698	34120	25619	40873
89756	12034	40138	62795
65981	43207	56480	17329
98563	01472	82071	95436
76849	20513	79362	01548

Figure 2.9

$y \equiv j$ are {ADG}, {BEH}, {CFI}. The first Latin square above has 0's in the $(0, 0)$, $(1, 2)$, and $(2, 1)$ entries, so the corresponding line is {AFH}. The remaining lines in this pencil are {BDI} and {CEG}, while the 0 entries in the second Latin square give the line {AEI}, the 1 entries give {BFG}, and the 2 entries give {CDH}. This is the plane of order 3 first described in Example 7.

Since there is no affine plane of order 10, there do not exist 9 mutually orthogonal Latin squares of order 10. In fact the value of $H(10)$ is not known. The first pair of orthogonal Latin squares of order 10 was constructed in 1959. However, it is currently not known if there are even three orthogonal Latin squares of order 10. Figure 2.9 shows two orthogonal Latin squares of order 10, taken from (Hall 1986, p. 227, reprinted by permission of John Wiley & Sons, Inc.).

As was earlier stated, there is always an affine plane of order p^k whenever p is a prime number and $k \geq 1$. Thus there are affine planes of orders 2, 3, 4, 5, 7, 8, 9, and 11. There is none of order 6 or order 10. It is unknown whether there is an affine plane of order 12, or whether there are any affine planes that are not of prime-power order.

A famous theorem in combinatorial theory, the Bruck-Ryser Theorem, gives some conditions for the *non*existence of some affine planes as follows:

■ **THEOREM 10. The Bruck-Ryser Theorem.** There is no affine plane of order n if n is of the form $4k + 1$ or $4k + 2$ unless n is the sum of two squares.

Any *prime* number of the form $4k + 1$ is the sum of two squares. In particular, there is no affine plane of order 6 or 21, but the case of $n = 10$ ($= 3^2 + 1^2$) is not covered by the theorem, nor is $26 = 5^2 + 1^2$.

Since the existence of an affine plane of order 12 is an open problem, it follows that $H(12)$ is unknown; it has been shown, however, that it is at least 5.

Partial Affine Planes

The existence of some (not necessarily $n - 1$) mutually orthogonal Latin squares leads to the construction of planes that satisfy the parallel axiom but do not necessarily have a line containing each pair of points. More precisely,

Definition. *A partial affine plane* is a set of *points* and *lines* with an incidence relation between them satisfying the following three axioms:

PA1. Two points P and Q can be on *at most* one line.

PA2. There are at least three lines through every point and at least two points on every line.

PA3. Given point P not on line ℓ, there is exactly one line m containing P and parallel to ℓ (where two lines are defined to be *parallel* if they have empty intersection).

For every n there is a partial affine plane of order n with two pencils of parallel lines. This can be constructed by writing the numbers 1 to n^2 in a square matrix, and taking the rows to be one pencil of parallel lines and the columns to be another. But more can be said.

■   **THEOREM 11.** For $k > 2$ there is a partial affine plane of order n with k pencils of parallel lines if and only if there are $k - 2$ mutually orthogonal Latin squares of order n.

Proof: The proof follows from the same construction as was done for Theorem 9.   ■

Corollary. The finite partial affine planes of order n having four pencils of parallel lines correspond to the pairs of orthogonal Latin squares A and B of order n.

Thus the open question stated above can be rephrased in terms of affine geometry: Is there a partial affine plane of order 10 with 5 pencils of parallel lines?

EXERCISES

1. Use the affine plane of order 4 given in Example 8 to construct three orthogonal Latin squares.

2. Use the orthogonal Latin squares A and B given below to construct a partial affine plane of order 5, having four pencils of parallel lines.

$$A = \begin{bmatrix} 0 & 1 & 2 & 3 & 4 \\ 1 & 2 & 3 & 4 & 0 \\ 2 & 3 & 4 & 0 & 1 \\ 3 & 4 & 0 & 1 & 2 \\ 4 & 0 & 1 & 2 & 3 \end{bmatrix}, \quad B = \begin{bmatrix} 4 & 0 & 1 & 2 & 3 \\ 3 & 4 & 0 & 1 & 2 \\ 2 & 3 & 4 & 0 & 1 \\ 1 & 2 & 3 & 4 & 0 \\ 0 & 1 & 2 & 3 & 4 \end{bmatrix}.$$

3. Find a pair of orthogonal Latin squares of order 3, different from the pair given in the text, and use the squares to construct an affine plane of order 3.

2.6 PROJECTIVE PLANES

With the discovery in the nineteenth century of the independence of Euclid's parallel postulate, attention became focused on geometries in which the parallel axiom fails to hold. Thus one of the main interests of geometers between about 1820 and 1920 was *projective geometry* in which there are no parallel lines. Although affine geometry is more appealing intuitively than projective geometry, projective geometry is, in a sense, more satisfying mathematically than affine geometry. For one thing, it often happens that several theorems from affine geometry will reduce to a single statement in projective geometry. Furthermore there is a *duality* in projective planes which, as we will see, means that every theorem has a *dual* theorem, whose proof is obtained "for free." More formally,

Definition. A *projective plane* is an ordered pair $(\mathcal{P}', \mathcal{L}')$, where $\mathcal{P}'$ is a nonempty set of elements called *points*, and $\mathcal{L}'$ is a nonempty collection of subsets of $\mathcal{P}'$ called *lines* satisfying the following axioms:

*P*1. If P and Q are distinct points, there is a unique line ℓ' such that $P \in \ell'$ and $Q \in \ell'$. (The line "through" P and Q will be written $\ell'(P, Q)$.)

*P*2. If ℓ' and m' are distinct lines in $\mathcal{L}'$, then $\ell' \cap m' \neq \varnothing$.

*P*3. There are at least three points on each line; there are at least two lines.

By *P*1, any two lines can intersect in at most one point; thus *P*2 implies that any two lines intersect in exactly one point. Furthermore

■ **THEOREM 12.** If $(\mathcal{P}', \mathcal{L}')$ is any projective plane, then

 i. $\mathcal{L}'$ contains at least three lines,

 ii. Each point in $\mathcal{P}'$ is contained in at least three lines.

Proof: (i) Let ℓ' and m' be lines in $\mathcal{L}'$. Each of ℓ' and m' has at least three points (one point being common to both lines). Thus there is a point A on ℓ' but not m' and a point B on m' but not ℓ'. The line $\ell'(A, B)$ is different from ℓ' and m'. Thus $\mathcal{L}'$ contains at least three lines.

(ii) Let A be an arbitrary point, and let ℓ', m', and k' be three distinct lines. If A is on each of the lines, we are done. Otherwise, assume that A is not on ℓ', and let B, C, and D be distinct points on ℓ'. Then $\ell'(A, B)$, $\ell'(A, C)$, and $\ell'(A, D)$ are three lines containing A.   ■

Actually, the seven-point, seven-line Fano plane, with three points on each of its lines as described in Example 5, is the smallest projective plane. The proof is left as an exercise.

The next two projective planes are based on the real vector space $\mathbb{R}^3$ and the corresponding three-dimensional Euclidean geometry.

☐**EXAMPLE 9: The Euclidean hemisphere.** Let $\mathscr{P}' = \{(x, y, z) \in \mathbb{R} \times \mathbb{R} \times \mathbb{R}: x^2 + y^2 + z^2 = 1, z \geq 0,$ and $z = 0$ implies that $y > 0$ or $x = 1\}$ (i.e., $\mathscr{P}'$ is the set of points on the unit hemisphere whose base is exactly half of the points on the unit circle in the xy-plane. Let $\mathscr{L}'$ be those semicircles on $\mathscr{P}'$ that have center $(0, 0, 0)$ and radius 1. Note that $\mathscr{L}'$ consists of arcs on the hemisphere, each of which is the intersection of the hemisphere with a *plane* through the origin $(0, 0, 0)$. Since any two points at a distance 1 from the origin uniquely determines a circle whose center is the origin and radius is 1, two points in $\mathscr{P}'$ lie on one and only one line in $\mathscr{L}'$. In Euclidean 3-space, if two planes intersect, they intersect in a line. Consider two lines in $\mathscr{L}'$, each of which is the intersection of a plane with $\mathscr{P}'$. The two planes in question each contain the origin, hence intersect in a Euclidean line through the origin. This line contains exactly one point of the hemisphere, and this point is on the two original half circles (lines in $\mathscr{L}'$). Axiom $P3$ clearly holds, so the hemisphere is an example of a projective plane.   ☐

☐**EXAMPLE 10: The real projective plane.** An alternative approach to Example 9 is this. Let $\mathbb{R}^3 = \mathbb{R} \times \mathbb{R} \times \mathbb{R}$ be the three-dimensional real vector space. Define an equivalence relation $\sim$ on the nonzero elements of $\mathbb{R}^3$ so that $(x, y, z) \sim (x', y', z')$ if and only if there is a nonzero element r in $\mathbb{R}$ with $(x', y', z') = r(x, y, z)$.

Let $\mathscr{P}'$ be the set of equivalence classes of elements of $\mathbb{R}^3$. Note that an equivalence class, or *projective point*, consists of some nonzero vector and all its nonzero scalar multiples. Denote the equivalence class containing the vector (p, q, r) as $\langle p, q, r \rangle$; p, q and r are called *homogeneous coordinates* for the appropriate point in the real projective plane. An equivalence class clearly consists of the nonzero points on some Euclidean *line* through the origin of $\mathbb{R}^3$, and it intersects the unit hemisphere of Example 9 in exactly one point. Thus the points of the hemisphere are in one-one correspondence with the equivalence classes of nonzero elements of $\mathbb{R}^3$.

Let $\ell' \in \mathscr{L}'$ when ℓ' is the set of projective points $\langle p, q, r \rangle$ that satisfy a *homogeneous* equation of the form

$$ax + by + cz = 0.$$

Recall that $ax + by + cz = 0$ is the equation for a *plane* through $(0, 0, 0)$ in $\mathbb{R}^3$. Given two projective points, one can, by solving simultaneous equations, obtain a unique (up to multiplication by a nonzero scalar) homogeneous equation which they satisfy. Similarly, given homogeneous equations

$$ax + by + cz = 0$$

and

$$dx + ey + fz = 0$$

that are not scalar multiples of each other, there is a unique projective point in which they intersect.

Thus Axioms $P1$ and $P2$ for projective planes hold in this example. It is clear that there are at least three points on each line and that not all points are collinear. Thus $(\mathscr{P}', \mathscr{L}')$ is a projective plane.

Any affine plane can be enlarged to a projective plane by adjoining a common new point to each pencil of parallel lines and gathering the new points together to form an additional line, as described below.

Suppose that $(\mathscr{P}, \mathscr{L})$ is an affine plane. For each pencil Φ of parallel lines, define a symbol X_Φ. The symbols X_Φ are called *points at infinity*. Let $\mathscr{P}' = \mathscr{P} \cup \{X_\Phi : \Phi$ a pencil of parallel lines in $\mathscr{L}\}$. For each $\ell \in \mathscr{L}$, define $\ell' = \ell \cup \{X_\Phi\}$, where Φ is the unique pencil containing ℓ. Let $\ell_\infty = \{X_\Phi : \Phi$ a pencil of parallel lines in $\mathscr{L}\}$. Finally, define $\mathscr{L}' = \{\ell' : \ell \in \mathscr{L}\} \cup \{\ell_\infty\}$.

■ **THEOREM 13.** $(\mathscr{P}', \mathscr{L}')$ is a projective plane whenever $(\mathscr{P}, \mathscr{L})$ is an affine plane.

Proof: To prove Axiom $P1$, first consider P and Q points in $\mathscr{P}$. There is a unique affine line ℓ containing P and Q, therefore the line ℓ' is the unique projective line containing them. For points $P \in \mathscr{P}$ and X_Φ, P is in a unique affine line ℓ in pencil Φ, so the unique projective line containing P and X_Φ is ℓ'. The unique line containing the two points X_Φ and X_Γ is ℓ_∞.

Axiom $P2$ follows since lines of the form ℓ' and m' were either formed by augmenting intersecting affine lines ℓ and m, in which case they still intersect as projective lines, or they were formed from parallel lines ℓ and m, contained in some pencil Φ. Thus the lines ℓ' and m' intersect in X_Φ. The line ℓ' contains X_Φ where Φ is the pencil containing ℓ; thus $\ell' \cap \ell_\infty = X_\Phi$.

Since there are at least two points on every affine line ℓ, there are at least three points on every projective line ℓ'. Since every affine plane contains at least three pencils of parallel lines, ℓ_∞ be contains at least three points. Since there are at least two lines in every affine plane, there are at least two lines in $\mathscr{L}'$. Thus $P3$ holds, and $(\mathscr{P}', \mathscr{L}')$ is a projective plane. ■

The construction given in Theorem 13 can be reversed; removing a line and all its points from a projective plane results in an affine plane. For $(\mathscr{P}', \mathscr{L}')$ a projective plane, select any one of its lines and call that line ℓ_∞. Define $\mathscr{P} = \{P : P \in \mathscr{P}', P \notin \ell_\infty\} = \mathscr{P}' \setminus \ell_\infty$. For each line $\ell' \neq \ell_\infty$, define $\ell = \{P : P \in \ell', P \notin \ell_\infty\} = \ell' \setminus (\ell' \cap \ell_\infty)$. Let $\mathscr{L} = \{\ell : \ell' \in \mathscr{L}', \ell' \neq \ell_\infty\}$.

■ **THEOREM 14.** If $(\mathscr{P}', \mathscr{L}')$ is a projective plane; then, relative to any line ℓ_∞ in $\mathscr{L}'$, $(\mathscr{P}, \mathscr{L})$ is an affine plane.

Proof: Any two points of $\mathscr{P}'$ are in a unique projective line ℓ'. If the points remain in $\mathscr{P}$, they are still in line ℓ, and only in line ℓ, so Axiom $A1$ for affine planes holds. Suppose that $P \in \mathscr{P}$, and that P is not on the affine line ℓ. Suppose that $\ell' \cap \ell_\infty = Q \in \mathscr{P}'$. Let $m' = \ell'(P, Q)$. Then $\ell' \cap m' = Q$. Thus $\ell \cap m = \varnothing$, and m is the unique line containing P and parallel to ℓ. Since each line in $\mathscr{L}'$ contains at least three points, each line in $\mathscr{L}$ contains at least two points. Since there are at least three lines in $\mathscr{L}'$, there are at least two lines remaining in $\mathscr{L}$. Thus $(\mathscr{P}, \mathscr{L})$ is an affine plane.   ■

The correspondence between projective and affine planes shown in Theorems 13 and 14 leads to combinatorial results for finite projective planes as well.

If $(\mathscr{P}, \mathscr{L})$ is a *finite* affine plane of order n, then since it has $n + 1$ pencils of parallel lines, $n + 1$ points at infinity are adjoined to its n^2 points to form a projective plane; thus the corresponding projective plane $(\mathscr{P}', \mathscr{L}')$ has $n^2 + n + 1$ points. One new line at infinity is added to the $n^2 + n$ lines of $(\mathscr{P}, \mathscr{L})$; thus $(\mathscr{P}', \mathscr{L}')$ has $n^2 + n + 1$ lines. Each point $X_\Phi \in \ell_\infty$ is also on ℓ' for each of the n lines $\ell \in \Phi$; thus each X_Φ is on $n + 1$ lines. The points in $\mathscr{P}$ remain on $n + 1$ lines as well. Thus the following definition is the natural one for the order of a projective plane.

Definition. A finite projective plane is said to have *order n* if it has $n^2 + n + 1$ points (and therefore $n^2 + n + 1$ lines).

Duality

In Theorem 12 it was shown that every projective plane contains at least three lines. Thus the axioms for a projective plane are equivalent to the following statements:

P1′. Any two distinct points are contained in one and only one common line.

P2′. Any two distinct lines contain one and only one point in common.

P3′. There are at least three points on each line; there are at least three lines through each point.

Interchanging the words "point" and "line," "contains" and "is contained in," does not change the axioms. Therefore, for each statement about a projective plane, there is a *dual* statement formed by interchanging the words above. A given statement is true for a projective plane if and only if its dual is true. It follows that a finite projective plane has the same number of points as lines, and the same number of points on each line as lines through each point. Collinear points and concurrent lines are dual concepts, while a *triangle* (comprised of three noncollinear points and three nonconcurrent lines) is a self-dual projective figure.

In addition to dualizing statements about projective planes, each plane itself has a dual, namely the abstract system whose points are the lines of the original plane and whose lines are the points of the original plane. In $(\mathscr{L}', \mathscr{P}')$, ℓ is "contained in" P (or is *incident* with P) if and only if $P \in \ell$ in the plane $(\mathscr{P}', \mathscr{L}')$.

Definition. The projective planes $(\mathscr{P}', \mathscr{L}')$ and $(\mathscr{Q}', \mathscr{K}')$ are said to be *isomorphic* when there is a bijection $f: \mathscr{P}' \to \mathscr{Q}'$ such that whenever $A \in \ell'(B, C)$ in $(\mathscr{P}', \mathscr{L}')$, then $f(A) \in \ell'(f(B), f(C))$ in $(\mathscr{Q}', \mathscr{K}')$. The mapping f is called a *collineation*.

A collineation between the affine planes $(\mathscr{P}, \mathscr{L})$ and $(\mathscr{Q}, \mathscr{K})$ is also defined as a bijection between the points that maps collinear points to collinear points (or equivalently, that maps lines to lines).

EXERCISES

1. Prove that the Fano plane is the smallest projective plane.

2. Show that the Fano plane can be obtained from the affine plane of order 2 described in Example 3 by adjoining a line at infinity as described in Theorem 13.

3. Use the construction in Theorem 13 to describe projective planes of orders 3 and 4 (i.e., planes with 13 points and 21 points, respectively).

4. By constructing examples, show that the three axioms for projective planes are independent.

5. Assuming Axioms $P1$ and $P3$ for projective planes, prove that Axiom $P2$ is equivalent to the following denial of the parallel axiom:

    $P2''$. Given a point P not on a line ℓ', there is no line containing P and parallel to ℓ'.

6. a. Prove directly from the axioms that there is a bijection between any two lines in a projective plane, regarding each line as a set of points.

    *b. Prove directly that there is a bijection between the set of points of any projective plane and its set of lines.

7. a. Prove that in every projective plane four points may be found, no three of which are collinear.

    b. State the dual of the theorem in (a).

8.   a. Define a collineation f from the Fano plane to itself such that $f(A) = B$.

b. Show that the Fano plane is self-dual by defining a collineation which maps A to {CDE}.

*c. Given any two points X and Y in the Fano plane, show that there is a collineation mapping X to Y.

9. a. Show that the real projective plane, as given in Example 10, is self-dual.

b. Show that the planes defined in Examples 9 and 10 are isomorphic.

10. Describe a collineation f from the nine-point twelve-line affine plane of Example 7 to itself such that the line {ABC} is mapped to {ADG}.

*11. A *difference set* in $\mathbb{Z}_m$ is a subset $S = \{a_1, \ldots, a_{n+1}\}$ such that for all $k \neq 0$ in $\mathbb{Z}_m$, there is a unique pair a_i, a_j in S such that $k \equiv a_i - a_j \pmod{m}$. In the following exercises assume that $S = \{a_1, \ldots, a_{n+1}\}$ is a difference set in $\mathbb{Z}_m$.

a. Show that $m = n^2 + n + 1$.

b. Define $\mathscr{P}' = \{P_1, \ldots, P_m\}$ and $\mathscr{L}' = \{\ell_1, \ldots, \ell_m\}$ and show that $(\mathscr{P}', \mathscr{L}')$ is a projective plane of order n where $P_i \in \ell_j$ if and only if $i + j \in S$.

c. Find a three-element difference set in $\mathbb{Z}_7$, and use it to construct a projective plane of order 2. Show that the plane is isomorphic to the Fano plane.

d. Construct a projective plane of order 4 from the difference set $\{0, 1, 4, 14, 16\}$ in $\mathbb{Z}_{21}$.

SUGGESTED READING

D. Betten, Die 12 Lateinischen Quadre der Ordung 6, *M. Helungen aus dem Math. Sem. Giessen* **163** (1984).

Kenneth P. Bogart, *Introductory Combinatorics*, London: Pitman, 1983.

R. C. Bose, E. T. Parker, and S. Shrikhande, Further results on the construction of mutually orthogonal Latin squares and the falsity of a conjecture of Euler, *Can. J. Math.* **12** (1960), 189–203.

R. H. Bruck and H. J. Ryser, The non-existence of certain finite projective planes, *Can. J. Math* **1** (1949), 88–93.

H. S. M. Coxeter, *Projective Geometry*, 2d ed., New York: Springer-Verlag, 1987.

J. Dénes and A. D. Keedwell, *Latin Squares and their Applications*, Guildford, England: University Presses, 1974.

D. J. Foulis and M. Munem, *After Calculus: Algebra*, San Francisco: Dellen, 1988.

Marshall Hall, *Combinatorial Theory*, 2d ed., Wiley Interscience, 1986.

G. Tarry, Le Probléme des 36 officiers, *C. R. Assoc. Av. Sci.* **1** (1900), 122–123.

E. T. Parker, Construction of some sets of mutually orthogonal Latin squares, *Proc. AMS* **10** (1959), 946–949.

CHAPTER 3

Desarguesian Affine Planes

The Greek geometers introduced "geometric algebra" by adding and multiplying lengths of segments in the Euclidean plane. Descartes formalized geometric algebra when he represented the synthetic Euclidean plane as the real coordinate plane. In particular, the points on the x- and y-axes are in one-one correspondence with the real numbers; these points can be added and multiplied geometrically. Addition is done by parallel displacement, while multiplication uses proportional segments. In this chapter we will define and investigate the properties of addition and multiplication in affine planes.

3.1 THE FUNDAMENTAL THEOREM

For a well-defined addition to satisfy the associative and commutative laws, and for a well-defined multiplication to satisfy the associative law, Desargues' Theorem (Theorems 13 and 14 of Chapter 1) is needed. For multiplication to be commutative, Pappus' Theorem (Theorem 12 of Chapter 1) is needed. The statements of the theorems of Desargues and Pappus are meaningful in an arbitrary affine plane because they only involve parallelism. The proofs of these results, as given in Chapter 1, however, rely on concepts (e.g., congruence and similar triangles) relevant to Euclidean geometry but meaningless in arbitrary affine planes.

Thus this chapter concerns affine planes in which Desargues' Theorem is true. Its aim is to define addition and multiplication of points on a given line, relative to fixed units O and I, and to prove the following theorem.

■ **THEOREM 1. The fundamental theorem of synthetic affine planes.** Relative to two fixed points, O and I, any line in a Desarguesian affine plane is a *division ring* in that its addition and multiplication satisfy the following properties:

Addition	Multiplication
Associative	Associative
O is the identity	I is the identity
All elements have inverses	All nonzero elements have inverses
Commutative	
Multiplication distributes over addition	

A division ring in which multiplication is commutative is a *field*, as defined in Chapter 1.

3.2 ADDITION ON LINES

An addition can be defined on the points of any line through any fixed point O, called the *origin*, and is defined *relative to this* O. In Chapter 4 the effect of selecting different origins will be investigated, and comparisons of sums of points relative to distinct origins will be made.

Definition of addition. In any affine plane $(\mathcal{P}, \mathcal{L})$ select and fix a line $\ell \in \mathcal{L}$, and a point O $\in \ell$. (See Fig. 3.1.) For points A and C in ℓ, to add A and C (relative to O):

1. Select B not in ℓ.
2. Let m be the line through B and parallel to ℓ.
3. Let k be the line through A parallel or equal to $\ell(O, B)$.
4. Let D $= m \cap k$.
5. Let n be the line through D parallel or equal to $\ell(B, C)$.
6. A + C is defined to be $n \cap \ell$.

It will be shown in Section 3.4 that the addition of points on ℓ is independent of the choice of the fixed point B.

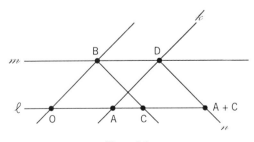

Figure 3.1

If A happens to be the special point O, then line k, as defined in step 3 is $\ell(O, B)$. The point D defined in step 4 always exists. Otherwise, m and k would be parallel, giving two lines (ℓ and k) through A and parallel to m, whereby $\ell = \mathit{k}$. But ℓ and k are distinct, for k is parallel or equal to $\ell(O, B)$, and ℓ is neither parallel nor equal to $\ell(O, B)$. Similarly n is not parallel to ℓ. Thus A + C as defined in step 6 exists.

Parallelograms

Parallelograms, familiar in Euclidean geometry, can also be defined in a general affine plane. A *parallelogram* is an ordered quadruple (A, B, C, D) of points such that $\ell(A, B) \| \ell(C, D)$ and $\ell(B, C) \| \ell(A, D)$. The lines $\ell(A, C)$ and $\ell(B, D)$ are called its *diagonals*. Two parallelograms arise in the preceding definition of addition of points, namely (O, A, D, B) and (C, A + C, D, B) which share the common vertices D and B. The definition of addition given here does, in fact, generalize addition on the (real) number line. In that case the *length* of the segment OA is added to the length of OC by moving OA by *parallel displacement*, first to BD and then to C(A + C).

■ **THEOREM 2.** O + C = C for any C in ℓ.

Proof: In the definition above A = O. Therefore, since $\mathit{k} = \ell(O, B)$ and B = D, we have $\mathit{n} = \ell(B, C)$ and O + C = C. ■

■ **THEOREM 3.** A + O = A for any A in ℓ.

Proof: Here $\ell(B, C) = \ell(B, O)$, so $\mathit{n} = \mathit{k}$. Since A $\in \mathit{k}$, $\mathit{k} \cap \ell = $ A, and A + O = A. ■

Either Theorem 2 or 3 implies the following result.

Corollary. O + O = O.

Clearly the fixed origin O acts as a "zero" or additive identity. However, in order that addition satisfy various other familiar properties such as commutativity and associativity, we need to assume *as an axiom* that Desargues' Theorem holds. This is the subject of the next section.

EXERCISES

1. Let $(\mathscr{P}, \mathscr{L})$ be the affine plane of order 3 defined in Example 7 of Chapter 2. On line {GHI} set G = O, and add H + I, using B as the point B was used in the definition of addition. Now compute H + I using whatever point you please for B.

2. Let $(\mathcal{P}, \mathcal{L})$ be the affine plane of order 3 as in Exercise 1. Make an addition table for {AEI}, setting A = O.

+	O	E	I
O			
E			
I			

3. Suppose that A, B, and C are noncollinear points in an affine plane $(\mathcal{P}, \mathcal{L})$. Prove that there is a unique point D in $\mathcal{P}$ such that (A, B, C, D) is a parallelogram.

4. Let $(\mathcal{P}, \mathcal{L})$ be the real Euclidean plane described in Example 1 of Chapter 2. Prove that $((a, b), (c, d), (e, f), (g, h))$ is a parallelogram if and only if $(a + e, b + f) = (c + g, d + h)$.

5. Make an addition table for the line {OAJH} in the plane of order 4, given in Example 8 of Chapter 2.

3.3 DESARGUES' THEOREM

In 1902 F. R. Moulton gave a simple example of an affine plane $(\mathcal{P}, \mathcal{L})$ called the *Moulton Plane*, in which Desargues' Theorem fails to hold. In this plane $\mathcal{P}$ is the set of ordered pairs of real numbers (i.e., the points of the real coordinate plane). $\mathcal{L}$ consists of all Euclidean horizontal lines, vertical lines, and lines with negative slope, together with all "broken"

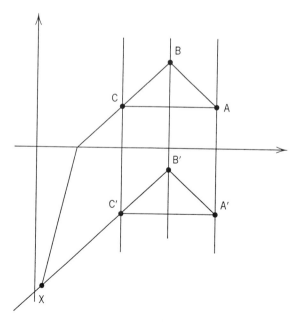

Figure 3.2

Euclidean lines that have positive slope m above the x-axis and slope $2m$ below the axis. It is an exercise in analytic geometry to show that any two points of the Moulton plane lie on one and only one (Moulton) line by finding its equation if it is a Euclidean line, or the point of intersection of the line with the x-axis if it is a broken line. Horizontal lines are parallel, vertical lines are parallel, and lines of the same negative slope are parallel. Broken lines whose slopes agree both above and below the x-axis are parallel as well, so it is apparent that the parallel axiom holds in the Moulton plane. Axiom $A3$ clearly holds. The easiest way to see that the Moulton plane does not satisfy Desargues' Theorem is to look at the example illustrated in Fig. 3.2 where the hypotheses of Desargues' Theorem are satisfied, but $\ell(B, C) \cap \ell(B', C') = X$.

Thanks to Moulton's example and others, it is clear that if Desargues' Theorem is to be used in affine plane geometry, it must be assumed as an additional axiom. Thus we make the following definition

Definition. A *Desarguesian affine* plane is an affine plane in which the following axiom holds (see Fig. 3.3):

$A4$. Desargues' Theorem.
 (I) Suppose that $\ell(A, A')\|\ell(B, B')\|\ell(C, C')$. If $\ell(A, B)\|\ell(A', B')$, and $\ell(A, C)\|\ell(A', C')$, then $\ell(B, C)\|\ell(B', C')$.
 (II) Suppose $\ell(A, A') \cap \ell(B, B') \cap \ell(C, C') = P$. If $\ell(A, B)\|\ell(A', B')$ and $\ell(A, C)\|\ell(A', C')$, then $\ell(B, C)\|\ell(B', C')$.

The rest of this chapter treats Desarguesian planes, while in Chapter 5 we will give some examples of finite non-Desarguesian planes.

Desargues' Theorem is named for the French mathematician and architect Girard Desargues (1593–1662), who first proved it. A contemporary and friend of Descartes, he also designed part of the current city hall in Lyon, France. As an affine axiom, Desargues' Theorem differs from

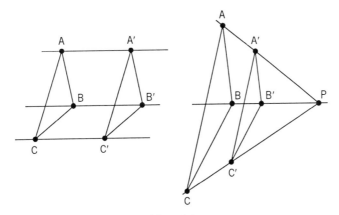

Figure 3.3

the first three in several ways. Clearly its statement is more complicated than the statements of the first three axioms, and in fact has two parts. Additionally it is equivalent to its own converse, which will be shown in Section 3.5. In Chapter 1 Desargues' Theorem was proved synthetically in the Euclidean plane; the following theorem shows that it holds in the coordinate real and rational planes.

■ **THEOREM 4.** The real and rational coordinate planes are Desarguesian.

Proof of (I): First note that A, A', B, for example are vectors (a_1, a_2), (a_1', a_2'), (b_1, b_2). (See Fig. 3.4): Since (A, C, C', A') is a parallelogram, $C' = C - A + A'$. Since (A, B, B', A') is a parallelogram, $B' = B - A + A'$. Thus $C' = C - B + B'$, so (C, B, B', C') is a parallelogram, with $\ell(B, C) \| \ell(B', C')$.

Proof of (II): If $P = (0, 0)$, then for some nonzero number m, $A' = mA$, $B' = mB$, and $C' = mC$. (See Fig. 3.4). Thus $B' - C' = m(B - C)$, so $\ell(B, C) \| \ell(B', C')$. If $P = (p_1, p_2)$, then subtracting the vector P from each of the other six vectors translates the configuration to one in which the point of intersection of the three lines is the origin. Thus $\ell(B - P, C - P) \| \ell(B' - P, C' - P)$. Adding back the vector P, we have $\ell(B, C) \| \ell(B', C')$. ■

The two statements of Desargues' Theorem for affine planes reduce to just one statement for projective planes. If in an affine plane $(\mathscr{P}, \mathscr{L})$ we have $\ell(A, A') \| \ell(B, B') \| \ell(C, C')$ or $\ell(A, A') \cap \ell(B, B') \cap \ell(C, C') = P$, after adjoining a line at infinity to form the projective plane $(\mathscr{P}', \mathscr{L}')$ the three lines meet at a point, possibly a point at infinity. Thus Axiom *P4*,

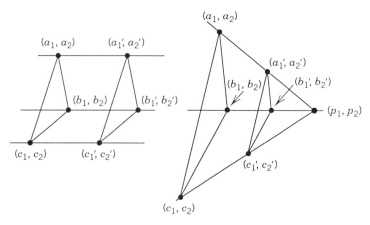

Figure 3.4

which follows, is the statement of Desargues' Theorem for projective planes.

Definition. A projective plane is said to be *Desarguesian* if it satisfies the following axiom (see Fig. 3.5):

*P*4. Suppose that $\ell(A, A') \cap \ell(B, B') \cap \ell(C, C') = P$. Let $C'' = \ell(A, B) \cap \ell(A', B')$, $B'' = \ell(A, C) \cap \ell(A', C')$, and $A'' = \ell(B, C) \cap \ell(B', C')$. Then A'', B'', and C'' are collinear.

Taking $\ell(A'', B'')$ as the line at infinity, gives Desargues' Theorem in the resulting affine plane, and the two affine statements of Desargues' Theorem correspond to the cases where P either is, or is not, on $\ell(A'', B'')$. Thus, the statement of Desargues' Theorem for the projective plane is an

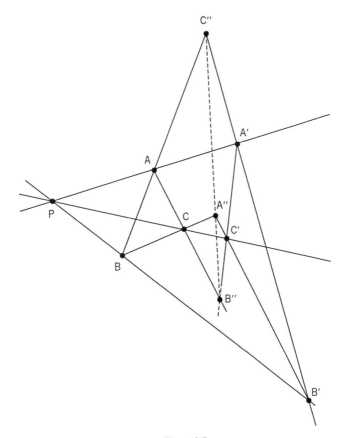

Figure 3.5

example of a single statement in projective geometry being equivalent to more than one affine statement.

Definition. A *triangle* (A, B, C) in a projective plane is a set of three noncollinear points called *vertices*, and the lines containing them called *sides*. The triangles (A, B, C) and (A', B', C') are called *perspective from a point* if the lines joining their vertices are *concurrent*, namely if $\ell(A, A') \cap \ell(B, B') \cap \ell(C, C') = P$ for some point; they are *perspective from a line* if their corresponding sides meet in collinear points. Thus Desargues' Theorem for the projective plane can be stated as follows:

P4'. If two triangles are perspective from a point, they are perspective from a line.

Note that "triangle" is a self-dual concept in a projective plane, while "perspective from a point" and "perspective from a line" are dual concepts. Therefore

■ **THEOREM 5.** A projective plane is Desarguesian if and only if it satisfies the converse of Desargues' Theorem.

EXERCISES

1. Find the point where the Moulton line ℓ containing $(2, 3)$ and $(-1, -4)$ crosses the x-axis. Find the point where the line containing $(3, 3)$ and parallel to ℓ crosses the x-axis.

2. Find an example of Desargues' configuration in the affine plane of order 3 (Example 7 of Chapter 2). Do the same for the affine plane of order 4 (Example 8 of Chapter 2).

3. State the converse of Desargues' Theorem as given in Axiom P4, and prove that it holds in any Desarguesian projective plane.

*4. Write Desargues' Theorem for affine planes as a single statement. State its converse.

5. Make sketches of the following configurations:
   a. ℓ and m are parallel lines with X, Y, Z, in ℓ and E, F, G, in m.
   b. Modify the sketch from (a) so that

$$\ell(Y, E) \| \ell(Z, F) \text{ and } \ell(Y, G) \| \ell(X, F).$$

6. Make sketches of the following configurations:
   a. (A, B, C, D) is a quadrilateral with no pair of its sides parallel.
   b. In the configuration from (a), insert A' so that (A', B, C, D) is a parallelogram.

  c. In the configuration from (b) insert B′ so that (A, B′, C, D) is a parallelogram.
  d. Similarly insert C′ and D′.

*7. Let $(\mathbb{R}, +, \circ)$ be the real numbers with their usual addition, and let k be a
  fixed positive real number different from 1. Define $r \circ s = rs$ if either r or s is
  positive or zero, and $r \circ s = krs$ if r and s are both negative. Let $(\mathscr{P}, \mathscr{L})$ be
  given by $\mathscr{P} = \{(x, y) : x, y \in \mathbb{R}\}$ and $\ell \in \mathscr{L}$ if and only if $\ell = \{(x, y) : a \circ x + b \circ y = c\}$ where a and b are not both zero.
  a. Show that $(\mathscr{P}, \mathscr{L})$ is an affine plane.
  b. For $k = 2$, sketch the line $1 \circ x + (-1) \circ y = 0$.
  c. For $k = 1/3$, sketch the line $1 \circ x + (-1) \circ y = 0$.
  d. For which value of k is $(\mathscr{P}, \mathscr{L})$ the Moulton Plane?

3.4 PROPERTIES OF ADDITION IN AFFINE PLANES

To define addition of coordinates on an affine line, two distinguished (or
"special") points were selected: the origin O on ℓ, and a point B (which
will be called an *auxiliary point*) not on ℓ. The effect of changing the O
will be considered later; now, using Desargues' Theorem, it can be shown
that regardless of the choice of B, the same sum (relative to O) will be
obtained for any fixed A and C in ℓ.

■   **THEOREM 6.** The definition of A + C is independent of the choice
of B.

  Proof: Add A and C, using two auxiliary points B_1 and B_2, obtaining
$(A + C)_1$ and $(A + C)_2$, as shown in Fig. 3.6. Note that once B_1 and B_2
are chosen, D_1 and D_2 are determined.)

  CASE 1. B_1, B_2, and O are collinear. The parallel axiom ensures that
D_1, D_2, and A are also collinear, on the line containing A and parallel to
$\ell(O, B_1)$. Now use Desargues' Theorem on triangles B_1, B_2, C and D_1,

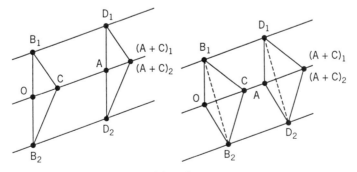

Figure 3.6

D_2, $(A + C)_1$, which gives $\ell(B_2, C)\|\ell(D_2, (A + C)_1)$. But we know that $\ell(B_2, C)\|\ell(D_2, (A + C)_2)$; this implies (by the parallel axiom and the fact that two lines intersect in at most one point) that $(A + C)_1 = (A + C)_2$.

CASE 2. $B_2 \notin \ell(O, B_1)$. Desargues' Theorem on triangles B_1, O, B_2 and D_1, A, D_2, gives $\ell(B_1, B_2)\|\ell(D_1, D_2)$. Then using Desargues' Theorem again, this time on triangles B_1, B_2, C and $D_1, D_2, (A + C)_1$, we obtain $\ell(B_2, C)\|\ell(D_2, (A + C)_1)$. As in Case 1, this leads to the conclusion that $(A + C)_1 = (A + C)_2$.   ■

Observe the use of the phrase "use Desargues' Theorem on triangles …" in the proof above. This is followed by two *ordered* triples of points, representing *corresponding* vertices. In the example above, the listing is B_1, B_2, C and $D_1, D_2, (A + C)_1$, where $\ell(B_1, D_1)$, $\ell(B_2, D_2)$, and $\ell(C, (A + C)_1)$ are parallel lines, while $\ell(B_1, B_2)\|\ell(D_1, D_2)$ and $\ell(B_1, C)\|\ell(D_1, (A + C)_1)$. It becomes much easier to use Desargues' Theorem if the six points involved are listed in the order given here.

In the sequel the phrase, "add $A + C$ using B (and D)" will also be used. This is almost self-explanatory. By putting "(and D)" in parentheses, one emphasizes the fact that choosing an *auxiliary* point B determines which point will play the role of D (the second auxiliary point) in the definition of addition.

Pursuing, as far as possible, the aim of imitating Descartes' introduction of *arithmetic* into geometry, it can now be shown that each point on ℓ has an *additive inverse* on ℓ.

■   **THEOREM 7.** For each point $A \in \ell$, there is a point $(-A) \in \ell$ such that $A + (-A) = (-A) + A = O$.

Proof: Select $B \notin \ell$, and let D be the point such that $\ell(B, D)\|\ell$ and $\ell(O, B)\|\ell(A, D)$ as in the definition of addition. Now let $(-A)$ be the point of intersection of ℓ with the line containing B and parallel to $\ell(O, D)$. (Check that this point exists!) Adding $A + (-A)$ using B (and D), we get $A + (-A) = O$. Adding $(-A) + A$ using D (and B), $(-A) + A = O$ as well. See Fig. 3.7 where this proof is illustrated.   ■

The proof that addition of points on a line is *associative* uses the fact that one has complete freedom in choosing "auxiliary" points for addition.

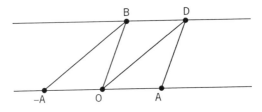

Figure 3.7

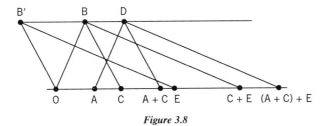

Figure 3.8

■ **THEOREM 8.** Given A, C, and E on ℓ, (A + C) + E = A + (C + E).

Proof: First add (A + C) using B (and D). Then add (A + C) + E using B' (chosen on ℓ(B, D) so that ℓ(O, B')$\|\ell$(B, C)). This will force D to be the second auxiliary point, and we obtain ℓ(B', E)$\|\ell$(D, (A + C) + E)). (see Fig. 3.8)

Now add (C + E) using B' (and B), and finally add A + (C + E) using B (and D). This gives ℓ(B, (C + E))$\|\ell$(D, A + (C + E)). Since ℓ(B', E)$\|\ell$(B, C + E), the parallel axiom implies that the points (A + C) + E and A + (C + E) are the same. ■

The last main theorem on addition states that addition is *commutative*. For this the *converse* of Desargues' Theorem for affine planes is needed. This, together with a proof that Desargues' Theorem (I) is implied by Desargues' Theorem (II) is the subject of the next section.

EXERCISES

1. Construct an addition table for the line {AEI} in the affine plane of order 3 (Example 7 in Chapter 2) with A = O. Use this table to find the inverse of the point E. Now find this inverse geometrically as in Theorem 7, and observe that they are the same.

   Exercises 2–4 refer to the affine plane of order 4 whose points and lines are repeated below. $\mathscr{P}$ = {A, B, . . . , P}, and the lines are grouped into pencils:

{OIDF}	{HBCD}	{HFKP}	{OAJH}	{HEIM}
{ECPA}	{EFGJ}	{BEOL}	{GIPB}	{BFAN}
{KMJB}	{IAKL}	{JNCI}	{FCML}	{CGKO}
{GHLN}	{MNOP}	{AGDM}	{EDKN}	{DJLP}

2. a. Find the line containing K and parallel to {BEOL}. Whree does this line intersect {MNOP}?

   b. On {MNOP} add M + N, using B as auxiliary point. What is the second auxiliary point?

3. Complete the addition table for {OIDF}. What is the inverse of D?

+	O	I	D	F
O				
I				
D				
F				

4. Find a parallelogram in the plane. What can you say about is diagonals?

5. Suppose that A, B, and C are noncollinear points in a Desarguesian affine plane $(\mathscr{P}, \mathscr{L})$. Let O be the unique point such that (A, B, C, O) is a parallelogram, and let D be the unique point such that (A, C, B, D) is a parallelogram. Relative to O, show that A + A = D.

6. Suppose that $(\mathscr{P}, \mathscr{L})$ is a Desarguesian affine plane in which the diagonals of every parallelogram are parallel. Suppose that A is any point in $\mathscr{P}$. Prove that relative to any O, A + A = O.

7. Find five orthogonal partitions of the numbers $\{1, \ldots, 16\}$.

8. Find three orthogonal Latin squares of order 4.

3.5 THE CONVERSE OF DESARGUES' THEOREM

As was stated above, Desargues' Theorem is equivalent to its own converse in the affine as well as the projective plane. By this is meant that the converse of Desargues' Theorem can be proved using Axioms $A1$, $A2$, $A3$, and $A4$, but also by assuming $A1$, $A2$, $A3$, and the converse of Desargues' Theorem, Desargues' Theorem can be proved. In fact more can be done. We will first use $A1$, $A2$, $A3$, and Desargues' Theorem (II) to prove the converse. Then we will use $A1$, $A2$, $A3$, and the converse to prove Desargues' Theorem. This will imply that, instead of assuming Desargues' Theorem, it would have been enough to assume only Desargues' Theorem (II).

■   **THEOREM 9. The converse of Desargues' Theorem for affine planes.** Let A, B, C and A', B', C' be two triples of noncollinear points in a Desarguesian affine plane. Suppose that

$$\ell(A, B) \| \ell(A', B'), \quad \ell(A, C) \| \ell(A', C'), \quad \text{and} \quad \ell(B, C) \| \ell(B', C').$$

Then either $\ell(A, A') \| \ell(B, B') \| \ell(C, C')$ or $\ell(A, A') \cap \ell(B, B') \cap \ell(C, C') = P$ for some P.

Proof (using A1, A2, A3, and Desargues' Theorem (II); see Fig. 3.9): Assume that $\ell(A, B) \| \ell(A', B')$; $\ell(A, C) \| \ell(A', C')$; $\ell(B, C) \| \ell(B', C')$. If it

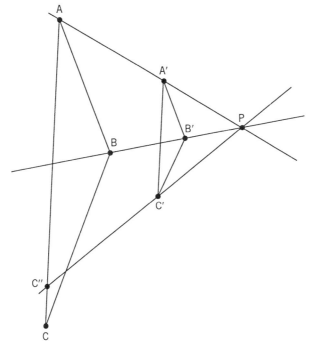

Figure 3.9

is not the case that the lines $\ell(A, A')$, $\ell(B, B')$, and $\ell(C, C')$ are all parallel, two of them must meet. Suppose that $\ell(A, A') \cap \ell(B, B') = P$. Let $\ell(C', P) \cap \ell(A, C) = C''$. Use Desargues' Theorem (II) on A, B, C'' and A', B', C' to obtain $\ell(B, C'')\|\ell(B', C')$. But $\ell(B, C)$ is the unique line through B and parallel to $\ell(B', C')$. Thus $\ell(B, C) = \ell(B, C'')$. This line intersects $\ell(A, C)$ in at most one point; hence $C = C''$, and $P \in \ell(C, C')$.
∎

■ **THEOREM 10.** Desargues' Theorem is true in any affine plane in which its converse holds.

Proof: (I) Suppose that A, B, C and A', B', C' are triples of noncollinear points. Suppose that $\ell(A, A')\|\ell(B, B')\|\ell(C, C')$, and let $\ell(A, B)\|\ell(A', B')$ with $\ell(A, C)\|\ell(A', C')$. It must be shown that $\ell(B, C)\|\ell(B', C')$.

Suppose not. Then let C'' be the point on $\ell(A, C)$ such that $\ell(B, C'')\|\ell(B', C')$. If $C \neq C''$, then $\ell(C'', C') \cap \ell(B, B') = P$ for some point P (they cannot be parallel since $\ell(C, C')$ is the unique line containing C'

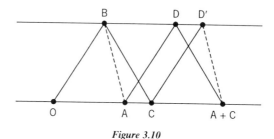

Figure 3.10

parallel to $\ell(B, B')$). Use the converse of Desargues' Theorem on A, B, C″ and A′, B′, C′ to get $\ell(A, A')$, $\ell(B, B')$ and $\ell(C'', C')$ all parallel or all meeting at one point. But the first two are assumed parallel, and the latter two intersect at P, a contradiction.

(II) Let A, B, C and A′, B′, C′ be triples of noncollinear points. Suppose $\ell(A, A') \cap \ell(B, B') \cap \ell(C, C') = P$, and let $\ell(A, B) \| \ell(A', B')$ with $\ell(A, C) \| \ell(A', C')$. It must be shown that $\ell(B, C) \| \ell(B', C')$.

Suppose not. Then let C″ be the point on $\ell(A, C)$ such that $\ell(B, C'') \| \ell(B', C')$. If $C \neq C''$, then P is not contained in $\ell(C'', C')$. Use the converse of Desargues' Theorem on A, B, C″ and A′, B′, C′ to get $\ell(A, A')$, $\ell(B, B')$, and $\ell(C'', C')$ all parallel or all meeting at one point. But the first two are assumed to meet at P, which is not in the third line, a contradiction. ■

Corollary. In any affine space in which Desargues' Theorem (II) holds, Desargues' Theorem holds as well.

■ **THEOREM 11.** Addition of points on a line in any Desarguesian affine plane is commutative.

Proof: Compute A + C using any point B (and D). Now find C + A using the same point B (and D′). Since $\ell(B, A) \| \ell(D', C + A)$, C + A is the same point as A + C if $\ell(B, A) \| \ell(D', A + C)$. (See Fig. 3.10)

First use the converse of Desargues' Theorem on A, D, A + C, and D′, C, B to show that the lines $\ell(A, D')$, $\ell(D, C)$, and $\ell(A + C, B)$ are all parallel or meet at a single point. Now use Desargues' Theorem on A, D, B, and D′, C, A + C to obtain $\ell(A, B) \| \ell(D', A + C)$, which completes the proof. ■

EXERCISES

1. Suppose that ℓ and m are parallel lines in a Desarguesian affine plane $(\mathscr{P}, \mathscr{L})$. Suppose that P, Q, and R are distinct points on ℓ, with S, T, and U distinct points on m as shown in Fig. 3.11. Assume that $\ell(P, S) \| \ell(R, U)$ and $\ell(P, T) \| \ell(Q, U)$. Prove that $\ell(Q, S) \| \ell(R, T)$.

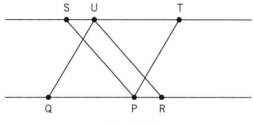

Figure 3.11

2. The statement in Exercise 1 is a theorem in Desarguesian affine planes. State the corresponding theorem for Desarguesian projective planes.

3. Let (A, B, C, D) be a quadrilateral in a Desarguesian affine plane, none of whose sides are parallel. Define A′ to be the unique point such that (A', B, C, D) is a parallelogram, and B′ the unique point such that (A, B', C, D) is a parallelogram as shown in Fig. 3.12. Prove that $\ell(A, A') \| \ell(B, B')$.

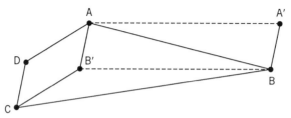

Figure 3.12

4. Let $(\mathscr{P}, \mathscr{L})$ be a Desarguesian affine plane containing the distinct points A, B, C, D, E, F, G. Assume that $\ell(A, D) \| \ell(B, G) \| \ell(E, F)$; $\ell(A, F) \| \ell(D, E)$;

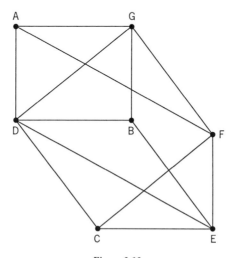

Figure 3.13

$\ell(A, G) \| \ell(D, B) \| \ell(C, E)$; $\ell(D, C) \| \ell(B, E)$ where all lines listed above are distinct. (See Fig. 3.13.) Prove:

a. $\ell(D, C) \| \ell(G, F)$.

b. $\ell(D, G) \| \ell(C, F)$.

3.6 MULTIPLICATION ON LINES OF AN AFFINE PLANE

In the preceding sections it was shown that addition is associative and commutative, O acts as an identity for addition, and every point has an additive inverse. These properties are discussed in Appendix A, together with an appropriate concept of *multiplication*, which can now be introduced for points on a line in an affine plane. Starting with a line ℓ, through a fixed origin O, as above, a fixed point I, *different* from O is chosen (recall that Axiom $A3$ guarantees at least two points on each line). Having these points, multiplication can be defined with respect to the selected O and I.

Definition of multiplication. Again $(\mathscr{P}, \mathscr{L})$ is a Desarguesian affine plane, with ℓ a line and O and I distinguished points of ℓ. (See Fig. 3.14) For points A and C in ℓ, to multiply A and C (relative to O and I):

1. Select B not in ℓ.
2. Let m be the line containing A, and parallel or equal to $\ell(I, B)$.
3. Let $D = m \cap \ell(O, B)$.
4. Let k be the line containing D and parallel or equal to $\ell(B, C)$.
5. AC is defined to be $k \cap \ell$.

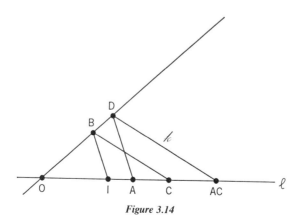

Figure 3.14

It will be shown in Theorem 15 that the multiplication of points on ℓ does not depend on the choice of the fixed point B.

Following the pattern established for addition, it is first shown that the suggestively named "I" acts as a multiplicative identity. Existence of nonzero inverses and associativity will follow as well; however multiplication is *not* in general commutative.

■ **THEOREM 12.** For any A in ℓ, AI = IA = A.

Proof: Select B (and D) as described above. Since $\ell(B, I)\|\ell(D, A)$, it follows that AI = A. In computing IA, note that line m described in step 2 is equal to $\ell(I, B)$, whence D = B and $k = \ell(B, A)$. Thus IA $(= \ell \cap k)$ = A.   ■

■ **THEOREM 13.** AO = OA = O for any A in ℓ.

Proof: In computing AO, the line k in step 4 is equal to $\ell(B, O)$, whence AO = O. In computing OA, the auxiliary point D is O, thus OA = O.   ■

■ **THEOREM 14.** If A ≠ O, there is a point (A^{-1}) in ℓ such that $A(A^{-1}) = (A^{-1})A = I$.

Proof: Select B and D so that $\ell(I, B)\|\ell(A, D)$ with D on $\ell(O, B)$. (See Fig. 3.15.) Let n be the line containing B and parallel to $\ell(D, I)$, and let $n \cap \ell = (A^{-1})$. Multiplying $A(A^{-1})$ using B and D gives I, and multiplying $(A^{-1})A$ using D and B also yields I.   ■

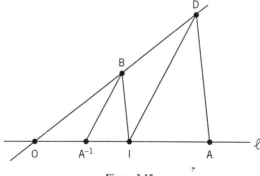

Figure 3.15

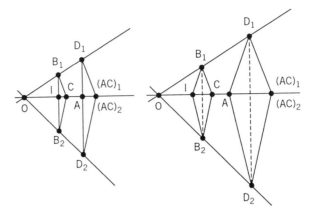

Figure 3.16

In Theorem 6 Desargues' Theorem (I) was used to prove that the sum of two points on a line (relative to a fixed origin O) does not depend on the choice of the auxiliary points. An analogous result (this time depending on Desargues' Theorem (II)) holds for multiplication. Observe also that as was the case with addition, the first auxiliary point determines the second one uniquely.

■   **THEOREM 15.** The definition of AC is independent of the choice of B above.

Proof: Obtain the product $(AC)_1$ by using B_1 (and D_1); then obtain $(AC)_2$ using B_2 (and D_2) as in Fig. 3.16.

CASE 1. I, B_1, and B_2 are collinear. Use Desargues' Theorem on B_1, B_2, C and D_1, D_2, $(AC)_1$ to obtain $\ell(B_2, C) \| \ell(D_2, (AC)_1)$, which forces $(AC)_1 = (AC)_2$.

CASE 2. The proof of Case 2 is left as an exercise.   ■

As was the case with addition, Theorem 15 can be used to show that multiplication of points on a line is associative.

■   **THEOREM 16.** For A, C, and E on ℓ, $(AC)E = A(CE)$.

Proof: (See Fig. 3.17, and compare this with the proof that addition is associative.) First multiply AC using B (and D). Now multiply (AC)E using B′ (and D), where $\ell(I, B') \| \ell(AC, D)$. Multiply CE using B′ (and B), and finally multiply A(CE) using B (and D) to obtain A(CE) = (AC)E.   ■

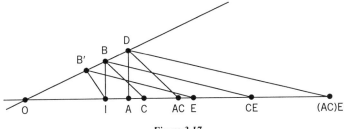

Figure 3.17

EXERCISES

1. In the affine plane of order 3 given in Example 7 of Chapter 2, fix ℓ as {CFI}, and let C = O. Make a multiplication table for the line (which is now {OIF}).
2. In the affine plane of order 4 given with the exercises of Section 3.4, p. 57, make a multiplication table for the line {OIDF}.

   a. What is the multiplicative inverse of D? of F?

   b. Compute (DD)F and D(DF) from the table, and check that they are the same.

   c. Compute D(I + F) and DI + DF from the table and check that they are the same.

3. Prove Case 2 of Theorem 15.
4. Let O, I, and A be distinct collinear points in a Desarguesian affine plane $(\mathscr{P}, \mathscr{L})$. Define the point 2I to be I + I and 2A to be the point A + A, the addition done relative to O. Prove geometrically that (2I)A = A + A assuming that the points O, I, 2I, A, and 2A are all distinct. [*Hint:* Compute I + I using B (and D) as auxiliary points as shown in Fig. 3.18. (2I)A is computed using B (and D′). Compute A + A using B and E.]

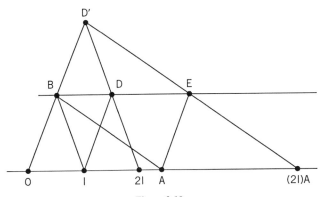

Figure 3.18

3.7 PAPPUS' THEOREM AND FURTHER PROPERTIES

Multiplication in a Desarguesian affine plane is not necessarily commutative. To present an example of a plane with noncommutative multiplication, some algebraic results are necessary, so the example is deferred until Chapter 4. However, the geometric counterpart of commutativity of multiplication is *Pappus' Theorem*, which was stated as a *Euclidean* theorem in Chapter 1, Theorem 12.

■   **PAPPUS' THEOREM.** Let $\ell \cap m = $ O, with P, Q, and R in ℓ; S, T, U in m; ℓ(P, T)∥ℓ(Q, U), and ℓ(Q, S)∥ℓ(R, T). Then ℓ(P, S)∥ℓ(R, U).

■   **THEOREM 17.** Pappus' Theorem is true in a Desarguesian affine plane if and only if multiplication on the points of its lines is commutative.

Proof: Choosing I in ℓ so that ℓ(I, S)∥ℓ(P, T), then PQ = R (multiplied using S (and T)). If Pappus' Theorem is true, then QP = R as well (multiplied using S (and U)). (See Fig. 3.19) Conversely, if Pappus' Theorem fails, then, relative to O and I, PQ = R but QP ≠ R, so multiplication is not commutative.   ■

Pappus and Desargues

In all the examples of Desarguesian affine planes presented so far, Pappus' Theorem holds. In fact algebraic methods can be used to show that Pappus' Theorem is true in any *finite* Desarguesian affine plane, but there are (infinite) Desarguesian affine planes in which Pappus' Theorem fails. On the other hand, it has been shown that Pappus' Theorem implies that of Desargues (Hessenberg 1905).

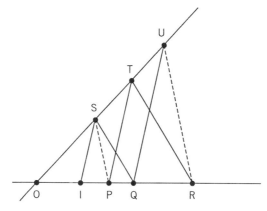

Figure 3.19

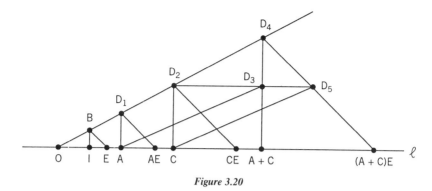

Figure 3.20

The Distributive Laws

Besides their individual properties, addition and multiplication are related by the *distributive laws*. Because multiplication is not necessarily commutative, two distributive laws are needed. Furthermore, after laws $(A + C)E = AE + CE$ and $E(I + C) = E + EC$ are proved *geometrically*, an algebraic proof shows that, for an arbitrary A, $E(A + C) = EA + EC$.

■   **THEOREM 18.** For A, C, and E in ℓ, $(A + C)E = AE + CE$.

Proof: Select $B \notin \ell$. (See Fig. 3.20.)
Compute:

AE using B (and D_1).
CE using B (and D_2).
A + C using D_2 (and D_3).
(A + C)E using B (and D_4).

Let D_5 be in $\ell(D_4, (A + C)E)$, so that $\ell(A, D_3) \| \ell(AE, D_5)$. Use Desargues' Theorem on D_4, D_3, D_5 and D_1, A, AE to get $\ell(D_3, D_5) \| \ell(A, AE)$, which shows that $D_5 \in \ell(D_2, D_3)$. Now add AE + CE using D_2 (and D_5), and note that $\ell(D_2, CE) \| \ell(D_5, AE + CE)$. But this means that AE + CE = (A + C)E.   ■

■   **THEOREM 19.** For C and E in ℓ, $E(I + C) = E + EC$.

Proof: Select $B \notin \ell$. (See Fig. 3.21.)
Compute:

EC using B (and D).
I + C using B (and D_1).
E + EC using D_1 (and D_2).

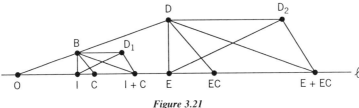

Figure 3.21

Use the converse of Desargues' Theorem on D_1, B, I and D_2, D, E which implies that $O \in \angle(D_1, D_2)$. Now use Desargues' Theorem on D_1, D_2, E + EC, and B, D_1, I + C, which gives $\angle(D, E + EC) \| \angle(B, I + C)$, whence E + EC = E(I + C). ∎

Corollary. For A, C, and E in $\angle$, E(A + C) = EA + EC.

Proof: If A = O, the result is immediate. If A ≠ O, then

$$\big(E(A + C) - EA - EC\big)A^{-1} = E\big((A + C)A^{-1} - E - ECA^{-1}$$

$$= E\big(I + CA^{-1}\big) - E - ECA^{-1}$$

$$= E + ECA^{-1} - E - ECA^{-1}$$

$$= O.$$

Thus E(A + C) − EA − EC = O, and

$$E(A + C) = EA + EC.$$ ∎

The fundamental theorem of synthetic affine geometry, stated as Theorem 1 of this chapter and repeated below, has now been proved.

∎ **THEOREM 20. The fundamental theorem of synthetic affine geometry.** Relative to fixed coordinates O and I, any line in a Desarguesian affine plane is a *division ring* in that its addition and multiplication are associative, and have identites, every element has an additive inverse, and every nonzero element has a multiplicative inverse. Addition is commutative, and multiplication distributes over addition.

In the next chapter we will prove that for any division ring $\mathbb{D}$, the elements of $\mathbb{D} \times \mathbb{D}$ are the points of a Desarguesian affine plane whose lines satisfy linear equations of the form $xa + yb = c$. Conversely, any Desarguesian affine plane can be so represented over (or *coordinatized by*) a division ring.

This chapter concludes with another example of an affine plane of order 5. It will be shown in Chapter 4 that this plane can be coordinatized by the field $\mathbb{Z}_5$.

☐ **EXAMPLE 1: An affine plane of order 5.** $\mathscr{P} = \{A, B, C, \ldots, Y\}$, and the lines are

{OIABC}	{OHMRV}
{DEFGH}	{IDNSW}
{JKLMN}	{AEJTX}
{PQRST}	{BFKPY}
{UVWXY}	{CGLQU}
{ODJPU}	{OGKTW}
{IEKQV}	{IHLPX}
{AFLRW}	{ADMQY}
{BGMSX}	{BENRU}
{CHNTY}	{CFJSV}
{OELSY}	{OFNQX}
{CDKRX}	{IGJRY}
{BHJQW}	{AHKSU}
{AGNPV}	{BDLTV}
{IFMTU}	{CEMPW}

☐

EXERCISES

The following exercises pertain to the plane of order 5 given above.

1. Make addition and multiplication tables for {OIABC}.
2. Make addition and multiplication tables for $\mathbb{Z}_5$. Show that {OIABC} is *isomorphic* to $\mathbb{Z}_5$ by finding a bijection from {OIABC} to $\mathbb{Z}_5$ that preserves addition and multiplication. [*Hint:* Map O to 0 and I to 1.]
3. Find a parallelogram in $(\mathscr{P}, \mathscr{L})$ and find the point of intersection of its diagonals.
4. Find an instance of the configurations for Desargues' Theorem in $(\mathscr{P}, \mathscr{L})$.
5. Find an instance of the configuration for Pappus' Theorem in $(\mathscr{P}, \mathscr{L})$.
6. Use $(\mathscr{P}, \mathscr{L})$ to define a projective plane of order 5.
*7. Let $(\mathscr{P}', \mathscr{L}')$ be a projective plane in which Pappus' Theorem holds, and prove Desargues' Theorem as follows: Let $\ell(A, A') \cap \ell(B, B') \cap \ell(C, C') = O$, and let $A'' = \ell(B, C) \cap \ell(B', C')$, $B'' = \ell(A, C) \cap \ell(A', C')$, and $C'' = \ell(A, B) \cap \ell(A', B')$. Define $E = \ell(A, C) \cap \ell(B', C')$, $F = \ell(A, B') \cap \ell(O, C)$, $G = \ell(A, B) \cap \ell(O, E)$, and $H = \ell(A', B') \cap \ell(O, E)$. Prove that A'', B'', and C'' are collinear by using Pappus' Theorem on O, B, B' and A, E, C; then on O, A, A' and B', C', E, and finally, on A, B', F and H, G, E. [*Hint:* In the first use of Pappus' Theorem, since $\ell(B, C) \cap \ell(B', E) = A''$, $\ell(A, B') \cap \ell(O, C) = F$ and $\ell(O, E) \cap \ell(A, B) = G$, we obtain A'', F, and G collinear.]

8. Find six orthogonal partitions of the set $\{1, 2, \ldots, 25\}$.

9. Find four orthogonal Latin squares of order 5.

SUGGESTED READING

G. Hessenberg, Beweis des Desarguesschen Satzes aus dem Pascalschen, *Math. Ann.* **61** (1905), 161–172.

David Hilbert, *Foundations of Geometry*, La Salle, IL: Open Court, 1971.

F. R. Moulton, A simple non-Desarguesian plane geometry, *Trans. AMS* **3** (1902), 192–195.

CHAPTER 4

Introducing Coordinates

On each line of any Desarguesian affine plane an addition and a multiplication can be defined, making the points of the line into a division ring. Furthermore, for any division ring $\mathbb{D}$, it will be shown in Section 4.3 that the coordinate plane $\mathbb{D}^2$ is a Desarguesian affine plane. The main purpose of this chapter is to show the converse: that every Desarguesian affine plane can be considered as a $\mathbb{D}^2$ upon renaming its points as ordered pairs of a ring elements and associating a linear equation with each line.

4.1 DIVISION RINGS

The properties of division rings, listed in the Fundamental Theorem of Synthetic Affine Geometry, are stated formally below.

Definition. A *division ring* is a set $\mathbb{D}$ with addition and multiplication that satisfy the following axioms:

$D1.$ $\mathbb{D}$ is closed under addition and multiplication.

$D2.$ Addition and multiplication are associative.

$D3.$ Addition is commutative.

$D4.$ $\mathbb{D}$ has an additive identity 0 and a two-sided multiplicative identity 1 such that $1 \times a = a \times 1 = a$ for every a in $\mathbb{D}$.

$D5.$ Every element in $\mathbb{D}$ has an additive inverse; every nonzero element a has a two-sided multiplicative inverse a^{-1} such that $a \times a^{-1} = a^{-1} \times a = 1$.

$D6.$ The distributive laws $a(b + c) = ab + ac$ and $(b + c)a = ba + ca$ hold in $\mathbb{D}$.

A *field*, as defined in Chapter 1, is a division ring in which multiplication is commutative.

Thus the Fundamental Theorem of Synthetic Affine Geometry may be restated as follows:

■ **THEOREM 1.** Let $(\mathscr{P}, \mathscr{L})$ be a Desarguesian affine plane, let $\ell \in \mathscr{L}$, and let O and I be any two distinct points on ℓ. Then $(\ell, +, \times)$ is a division ring where $+$ and $\times$ are defined relative to O and I. Furthermore $(\ell, +, \times)$ is a field if Pappus' Theorem holds in $(\mathscr{P}, \mathscr{L})$.

Proof: D1 follows from the definitions of addition and multiplication. D2 follows from Theorem 6 and 14 of Chapter 3.

D3 follows from Theorem 9.

D4 follows from Theorem 2, 3, and 10.

D5 follows from Theorems 5 and 12.

D6 follows from Theorems 18, 19, and 20.

By Theorem 17 of Chapter 3, Pappus' Theorem holds in $(\mathscr{P}, \mathscr{L})$ if and only if each $(\ell, +, \times)$ is a field. ■

Clearly every field is a division ring. Actually a remarkable theorem of J. H. M. Wedderburn (1882–1948) states that every *finite* division ring is a field. Thus the simplest example of a noncommutative division ring is infinite, and is presented below.

□**EXAMPLE 1:** The quaternions. $\mathbb{H}$ (after Sir William Rowan Hamilton, and Irish mathematician [1805–1865] who discovered them in 1853) is defined to be the set

$$\{a + bi + cj + dk : a, b, c, d \text{ real numbers}\}.$$

Addition and multiplication are defined by considering the quaternions as polynomials in i, j, and k. Therefore $(a + bi + cj + dk) + (a' + b'i + c'j + d'k) = (a + a') + (b + b')i + (c + c')j + (d + d')k$. Similarly multiplication is polynomial multiplication subject to the rules

$$i^2 = j^2 = k^2 = -1,$$

$$ij = -ji = k, \quad jk = -kj = i, \quad ki = -ik = j.$$

Thus the product of the two quaternions given above is $aa' - bb' - cc' - d' + (ab' + ba' + cd' - dc')i + (ac' + ca' + db' - d'b)j + (ad' + da' + be'cb')k$. It can also be verified that for a, b, c, and d not all 0, then $(a + bi + cj + dk) \times (1/(a^2 + b^2 + c^2 + d^2))(a - bi - cj - dk = 1 + 0i + 0j + 0k = 1$. The quaternions are not a field since, for example, $ij \neq ji$. (For further details on this approach, consult Birkhoff and MacLane 1977, p. 255.) □

An Alternative Approach to $\mathbb{H}$

The quaternions can also be constructed as the set of 2×2 complex matrices of the form

$$\begin{pmatrix} a + bi & c + di \\ -c + di & a - bi \end{pmatrix}.$$

The quaternions satisfy the properties $D1$ through $D6$ under the usual addition and multiplication of matrices. Matrix addition and multiplication are always associative and the distributive laws always hold.

The identity matrix,

$$\begin{pmatrix} 1 + 0i & 0 + 0i \\ 0 + 0i & 1 - 0i \end{pmatrix}$$

is a quaternion and is the multiplicative identity for $\mathbb{H}$. Since the determinant of

$$\begin{pmatrix} a + bi & c + di \\ -c + di & a - bi \end{pmatrix}$$

is $a^2 + b^2 + c^2 + d^2$ which is different from 0 if the matrix is nonzero, each nonzero quaternion has an inverse under matrix multiplication. Furthermore this inverse is again a quaternion (see Exercise 2c).

Almost all the division rings occurring frequently in mathematics are fields; in fact the quaternions will be the *only* example of a noncommutative division ring presented in this chapter. Although there are other (necessarily infinite) noncommutative division rings, $\mathbb{H}$ is the simplest such a ring and the best known. Interestingly enough, the next most common example was first presented, not in a volume on algebra, but by Hilbert in his *Foundations of Geometry*, where *geometric* and analytic, not algebraic, considerations led him to develop his (more complicated) example, which is given here in Appendix B. It should be noted that in 1899 when Hilbert wrote, the word "ring" had not been defined, and Hilbert referred to what are now called division rings as "complex number systems."

EXERCISES

1. Solve the equation $x(i + j) = 3 + k$ for $x \in \mathbb{H}$.
2. Consider the quaternions as complex 2×2 matrices in the form given in the text for the following exercises.
   a. Show that $\mathbb{H}$ is closed under matrix addition and multiplication.
   b. Show, by giving an example, that multiplication in $\mathbb{H}$ is not commutative.
   c. Show that the inverse of any nonzero quaternion is a quaternion.
3. Show that $\{\pm 1, \pm i, \pm j, \pm k\}$ is an eight-element non-abelian group (called the *quaternion group*) under quaternion multiplication.

4. Suppose that $\mathbb{D}$ satisfies all axioms for a division ring except the distributive laws. Suppose further that $a(b + c) = ab + ac$ for all a, b and c in $\mathbb{D}$ and that $(1 + c)a = a + ca$ for all a and c in $\mathbb{D}$. Prove that $\mathbb{D}$ is a division ring.

*5. Prove that the equation $x^2 = -1$ has an infinite number of solutions in $\mathbb{H}$.

*6. Suppose that $q = a + bi + cj + dk$ is a non-real quaternion. Prove there is a quaternion q' such that $qq' \neq q'q$.

4.2 ISOMORPHISM

In Chapter 2, it was shown that any two lines in an affine plane contain the same number of points. Furthermore each line in $\mathbb{R}^2$ is a real number line in the sense that its members correspond to the real numbers. This idea can be extended to algebraic systems by defining two division rings to be *isomorphic* if they have not only the same cardinality but the same mathematical properties. More formally,

Definition. Division rings $\mathbb{D}$ and $\mathbb{D}'$ are said to be *isomorphic* (written $\mathbb{D} \cong \mathbb{D}'$) if there is a one-one correspondence $\phi : \mathbb{D} \to \mathbb{D}'$ such that

$$\phi(a + b) = \phi(a) + \phi(b)$$

and

$$\phi(ab) = \phi(a)\phi(b)$$

whenever a and b are in $\mathbb{D}$. In this case the mapping ϕ is called an *isomorphism* from $\mathbb{D}$ to $\mathbb{D}'$. (Since every field is a division ring, two fields are isomorphic if they are isomorphic as division rings.)

Clearly any isomorphism preserves the cardinality of a division ring. But more can be said. The following theorem shows that any isomorphism of division rings maps identities to identities and inverses to inverses.

■   **THEOREM 2.** Let ϕ be an isomorphism from $\mathbb{D}$ to $\mathbb{D}'$. Then

i. $\phi(0) = 0$ (the first 0 being in $\mathbb{D}$, the second in $\mathbb{D}'$),
ii. $\phi(1) = 1$,
iii. $\phi(-a) = -\phi(a)$ for every a in $\mathbb{D}$,
iv. $\phi(a^{-1}) = (\phi(a))^{-1}$ for every nonzero a in $\mathbb{D}$.

Proof: (i) Let b be an element of $\mathbb{D}'$. Then, since ϕ is onto, $b = \phi(a)$ for some a in $\mathbb{D}$. Thus $b + \phi(0) = \phi(a) + \phi(0) = \phi(a + 0) = \phi(a) = b$, whence $\phi(0)$ is the additive identity of $\mathbb{D}'$.

(iii) For a in $\mathbb{D}$, $0 = \phi(0) = \phi(a + (-a)) = \phi(a) + \phi(-a)$. Thus (adding $-\phi(a)$ to both sides) gives $-\phi(a) = \phi(-a)$.

The proofs of parts ii and iv are similar and are left as exercises.   ■

It follows from Theorem 2 that isomorphic division rings have the same mathematical properties. For example, if $\mathbb{D} \cong \mathbb{D}'$ and $\mathbb{D}$ is commutative (i.e., is a field), so is $\mathbb{D}'$.

■ **THEOREM 3.** Composites of isomorphisms are again isomorphisms.

Proof: Let $\phi : \mathbb{D} \to \mathbb{D}'$ and $\sigma : \mathbb{D}'' \to \mathbb{D}$ be isomorphisms of division rings. Then $\phi \circ \sigma : \mathbb{D}'' \to \mathbb{D}'$ is a bijection by Theorem 1 of Chapter 2. Suppose that a and b are in $\mathbb{D}''$. Then $(\phi \circ \sigma)(a + b) = \phi(\sigma(a + b)) = \phi(\sigma(a) + \sigma(b)) = \phi(\sigma(a)) + \phi(\sigma(b)) = (\phi \circ \sigma)(a) + (\phi \circ \sigma)(b)$, and the composition preserves addition. The proof that it preserves multiplication is quite similar and is omitted.   ■

Suppose that two finite fields F and F' are isomorphic. Observe that upon renaming each element of F by its image in F', the addition and multiplication tables of F become identical to those of F'. Thus saying that F and F' are isomorphic fields is really saying that they are mathematically the *same* field, although the elements may have different names. This can be demonstrated by taking F and F' to be $\mathbb{Z}_5$ and the line {OIABC} from the affine plane of order 5 (see Exercise 6 below).

In Appendix A finite fields of orders p^2 and p^3 are constructed, where p is a prime number. Similar methods can be used to construct fields of orders p^k where k is any positive integer. It is a theorem of abstract algebra that any two fields of the same finite order are isomorphic, and that there is a field $GF(p^k)$ for every prime p and positive integer k (see Birkhoff and Mac Lane 1977, sec. 15.3). Conversely, every finite field has prime power order. Thus we can speak of "the" field of order p^k as $GF(p^k)$.

EXERCISES

The reader is advised to consult Appendix A, especially in doing Exercises 4 and 8.

1. Prove parts 2 and 4 of Theorem 2.

2. Complete the proof of Theorem 3 by showing that if $\phi : \mathbb{D} \to \mathbb{D}'$ and $\sigma : \mathbb{D}'' \to \mathbb{D}$ are isomorphisms, then $\phi \circ \sigma : \mathbb{D}'' \to \mathbb{D}'$ preserves multiplication.

3. Prove that complex conjugation defined by $\phi(a + bi) = a - bi$ is an isomorphism of $\mathbb{C}$.

4. Prove that the mapping ϕ on $GF(4) = \mathbb{Z}_2[u]/(u^2 + u + 1)$ described by $\phi(u) = u + 1$ is an isomorphism.

5. Show that the complex field $\mathbb{C}$ and the real field $\mathbb{R}$ are not isomorphic. [*Hint:* Consider the image of i under such an "isomorphism."]

6. Show that the line {OIABC} of the affine plane of order 5 given in Example 1 of Chapter 3, with addition and multiplication defined geometrically, is isomorphic to $\mathbb{Z}_5$. [*Hint:* Map 0 to O, 1, to I, 2 to I + I, etc.]

7. Show that the two representations for the quaternions given above (as matrices and as polynomials in i, j, and k) give isomorphic division rings.

8. Exhibit an isomorphism between the nine element fields $\mathbb{Z}_3[u]/(u^2 - u - 1)$ and $\mathbb{Z}_3[u]/(u^2 + 1)$.

4.3 COORDINATE AFFINE PLANES

Vector spaces can be formed over division rings as well as fields, as long as care is taken regarding the side on which scalar multiplication is performed.

Definition. Let $\mathbb{D}$ be a division ring and n a positive integer. Then the (*n-dimensional*) *left vector space over* $\mathbb{D}$ (written $\mathbb{D}^n$) is the set $\{(x_1, x_2, \ldots, x_n): x_i \in \mathbb{D}\}$. The elements of this set are called (*n*-dimensional) vectors, and when the context is clear, the vector $(x_1, x_2, \ldots, x_n)$ will be denoted $\mathbf{x}$. The vector $\mathbf{0} = (0, 0, \ldots, 0)$ is called the *origin* of $\mathbb{D}^n$.

The vector space $\mathbb{D}^n$ has two operations, componentwise addition and scalar multiplication, which are defined as follows:

Definition. Let $\mathbf{x}$ and $\mathbf{y}$ be in $\mathbb{D}^n$. Then $\mathbf{x} + \mathbf{y}$ is defined to be the vector $(x_1 + y_1, x_2 + y_2, \ldots, x_n + y_n)$. For a in $\mathbb{D}$ and $\mathbf{x}$ in $\mathbb{D}^n$, $a\mathbf{x}$ is defined to be $(ax_1, ax_2, \ldots, ax_n)$.

Since we will only deal with left vector spaces here (i.e., those in which scalars multiply on the left side of vectors), we will use the term "vector space" to describe them. Right vector spaces are defined analogously, but will not be used here. Many of the notions from $\mathbb{R}^n$ carry over to $\mathbb{D}^n$. Simultaneous equations, for example, can be solved in vector spaces over division rings just as in real vector spaces. We leave the proofs of the main properties of addition and scalar multiplication as exercises. The reader will note that the proofs valid in $\mathbb{R}^n$ carry over to $\mathbb{D}^n$.

We will prove in this section that for any division ring $\mathbb{D}$, $\mathbb{D}^2$ is a Desarguesian affine plane, and in Section 4.5 we will prove the more difficult converse, that any Desarguesian affine plane can be represented (coordinatized) as $\mathbb{D}^2$ for some division ring $\mathbb{D}$. As might be expected, the points of $\mathbb{D}^2$ are its vectors (x, y), and its lines satisfy linear equations. However, the linear equations must be written with coefficients on the *right* and thus are of the form $xa + yb = c$.

■ **THEOREM 4.** Let $\mathbb{D}$ be a division ring. Define $(\mathscr{P}, \mathscr{L})$ by letting $\mathscr{P} = \mathbb{D}^2$ and $\ell \in \mathscr{L}$ if and only if there are elements a, b, and c in $\mathbb{D}$, with a and b not both 0, such that $\ell = \{(x, y) \in \mathbb{D}^2: xa + yb = c\}$. Then $(\mathscr{P}, \mathscr{L})$ is an affine plane.

Remark: If $\mathbb{D}$ is a field, the linear equation given above can be written in the more familiar form $ax + by = c$. However, if $\mathbb{D}$ is noncommutative, there is a certain subtlety here; namely if *left* scalar multiplication is used in $\mathbb{D}^2$, the coefficients of the linear equations defining lines must be written on the *right*.

Proof of Axiom A1: Let (x_1, y_1) and (x_2, y_2) be in $\mathbb{D}^2$. We must find a, b, and c in $\mathbb{D}$, with a and b not both zero, such that $xa + yb = c$ when $(x,y) = (x_1, y_1)$ or (x_2, y_2).

If $x_1 = x_2$, then both (x_1, y_1) and (x_2, y_2) satisfy $x = x_1$. This corresponds to the *vertical* lines in the real coordinate plane.

If $x_1 \neq x_2$, we get the line with finite "slope" $\lambda = (x_2 - x_1)^{-1}(y_2 - y_1)$ and equation $y = y_1 + (x - x_1)\lambda$, whose intercepts are $(0, y_1 - x_1\lambda)$ and $(x_1 - y_1\lambda^{-1}, 0)$.

To show the uniqueness of the line, assume $x_1 \neq x_2$. We show that the following equations have a unique solution:

$$x_1 + y_1 b = c,$$

$$x_2 + y_2 b = c.$$

Subtracting (i.e., multiplying the second equation through by $-1 \in \mathbb{D}$ and then adding) gives

$$(x_1 - x_2) + (y_1 - y_2)b = 0.$$

Thus

$$b = (y_1 - y_2)^{-1}(x_2 - x_1)$$

and

$$c = x_1 + y_1(y_1 - y_2)^{-1}(x_2 - x_1).$$

Therefore b and c are uniquely determined. Any other linear equation satisfied by (x_1, y_1) and (x_2, y_2) is a scalar multiple of $x + yb = c$. If $x_1 = x_2$, then $y_1 \neq y_2$, and the proof proceeds as above.

Proof of Axiom A2: The parallel axiom is easy to verify after observing that, for c *different from* d in $\mathbb{D}$ the lines with equations

$$xa + yb = c$$

and

$$xa + yb = d$$

cannot have any point in common. For (x_1, y_1) to be on both lines implies that $c = x_1 a + y_1 b = d$, a contradiction. Thus, given (x_1, y_1) not in the line with equation $xa + yb = c$, the line containing that point and parallel to the line has equation $xa + yb = x_1 a + y_1 b$. The verification of the uniqueness of the line is left as an exercise.

Proof of Axiom A3: Since $\mathbb{D}$ has at least two *distinct* elements 0 and 1, the line given by $xa + yb = c$ contains at least the points $(ca^{-1}, 0)$ and $((c - b)a^{-1}, 1)$, assuming that $a \neq 0$. The lines with equations $x = 0$ and $x = 1$ are different, so there are at least two lines. ∎

The following three results on parallelism are needed to prove that Desargues' Theorem holds in $\mathbb{D}^2$.

■ **THEOREM 5.** For $\mathbf{u}, \mathbf{v}, \mathbf{z}$, and $\mathbf{w}$ in $\mathbb{D}^2$, $(\mathbf{u}, \mathbf{v}, \mathbf{w}, \mathbf{z})$ is a parallelogram if and only if $\mathbf{z} = \mathbf{u} - \mathbf{v} + \mathbf{w}$.

Proof: If the line containing vectors $\mathbf{u} = (u_1, u_2)$ and $\mathbf{v} = (v_1, v_2)$ in $\mathbb{D}^2$ satisfies the equation $xa + yb = c$, and $\mathbf{w} = (w_1, w_2)$ is not on that line, then $(\mathbf{u}, \mathbf{v}, \mathbf{w}, \mathbf{u} - \mathbf{v} + \mathbf{w})$ is a parallelogram. This follows from the fact that both $\mathbf{w}$ and $\mathbf{u} - \mathbf{v} + \mathbf{w}$ satisfy the equation

$$xa + yb = w_1 a + w_2 b.$$

Clearly the equation is satisfied by $\mathbf{w}$, but $(u_1 - v_1 + w_1)a + (u_2 - v_2 + w_2)b = u_1 a + u_2 b - v_1 a - v_2 b + w_1 a + w_2 b = c - c + w_1 a + w_2 b = w_1 a + w_2 b \neq c$. Similarly the line through $\mathbf{u}$ and $\mathbf{z}$ is parallel to the one through $\mathbf{v}$ and $\mathbf{u} - \mathbf{v} + \mathbf{w}$, so $(\mathbf{u}, \mathbf{v}, \mathbf{w}, \mathbf{u} - \mathbf{v} + \mathbf{w})$ is a parallelogram. By Axioms $A1$ and $A2$ proved above, $\mathbf{u} - \mathbf{v} + \mathbf{w}$ is the unique point such that $(\mathbf{u}, \mathbf{v}, \mathbf{w}, \mathbf{u} - \mathbf{v} + \mathbf{w})$ is a parallelogram. ∎

■ **THEOREM 6.** Suppose that $\mathbf{z}, \mathbf{u}$, and $\mathbf{v}$ are distinct collinear points in $\mathbb{D}^2$. Then there is an element t in $\mathbb{D}$ such that $\mathbf{v} = \mathbf{u} + t(\mathbf{u} - \mathbf{z})$.

Proof: If $z_2 = u_2$, then $\mathbf{z}, \mathbf{u}$, and $\mathbf{v}$ satisfy the equation $y = z_2$, and the t in question is $(v_1 - u_1)(u_1 - z_1)^{-1}$.

Otherwise, $\mathbf{z}, \mathbf{u}$, and $\mathbf{w}$ satisfy the equation

$$x + y(z_2 - u_2)^{-1}(u_1 - z_1) = u_1 + u_2(z_2 - u_2)^{-1}(u_1 - z_1).$$

In particular, $v_1 + v_2(z_2 - u_2)^{-1}(u_1 - z_1) = u_1 + u_2(z_2 - u_2)^{-1}(u_1 - z_1)$ so that $v_1 = u_1 + [(u_2 - v_2)(z_2 - u_2)^{-1}](u_1 - z_1)$. But $v_2 = u_2 + (u_2 - v_2)[z_2 - u_2)^{-1}(u_2 - z_2)] = u_2 + (u_2 - v_2)(-1)$, so the t here is $(u_2 - v_2)(z_2 - u_2)^{-1}$. ∎

■ **THEOREM 7.** If $\ell(\mathbf{u}, \mathbf{v}) \| \ell(\mathbf{z}, \mathbf{w})$ in $\mathbb{D}^2$, then $\mathbf{u} - \mathbf{v}$ is a nonzero multiple of $\mathbf{w} - \mathbf{z}$. Conversely, if $\mathbf{u}, \mathbf{v}$ and $\mathbf{z}$ are noncollinear, and $\mathbf{u} - \mathbf{v}$ is a nonzero multiple of $\mathbf{w} - \mathbf{z}$, then $\ell(\mathbf{u}, \mathbf{v}) \| \ell(\mathbf{z}, \mathbf{w})$.

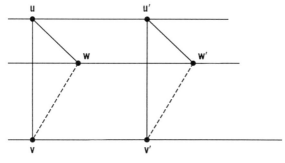

Figure 4.1

Proof: If $\mathbf{x}$ is the unique point in $\mathbb{D}^2$ such that $(\mathbf{u}, \mathbf{v}, \mathbf{z}, \mathbf{x})$ is a parallelogram, by Theorem 5 we have $\mathbf{x} = \mathbf{u} - \mathbf{v} + \mathbf{z}$, and by Theorem 6, $\mathbf{x} = \mathbf{z} + t(\mathbf{w} - \mathbf{z})$, where t is nonzero since $\mathbf{x} \neq \mathbf{z}$. Combining these, we have $\mathbf{u} - \mathbf{v} + \mathbf{z} = \mathbf{z} + t(\mathbf{w}) - \mathbf{z})$, so $\mathbf{u} - \mathbf{v} = t(\mathbf{w} - \mathbf{z})$.

Conversely, given $\mathbf{u}$, $\mathbf{v}$, and $\mathbf{z}$, $(\mathbf{u}, \mathbf{v}, \mathbf{z}, \mathbf{u} - \mathbf{v} + \mathbf{z})$ is a parallelogram. But $\mathbf{u} - \mathbf{v} + \mathbf{z} = \mathbf{z} + t(\mathbf{w} - \mathbf{z})$ and is therefore on the line $\ell(\mathbf{z}, \mathbf{w})$, so $\ell(\mathbf{u}, \mathbf{v}) \| \ell \mathbf{z}, \mathbf{w})$. ∎

■ **THEOREM 8. Desargues' Theorem I.** Suppose that $\mathbf{u}$, $\mathbf{v}$, and $\mathbf{w}$ are noncollinear points in $\mathbb{D}^2$. Then, if $(\mathbf{u}, \mathbf{u}', \mathbf{v}', \mathbf{v})$ and $(\mathbf{u}, \mathbf{u}', \mathbf{w}', \mathbf{w})$ are parallelograms, $(\mathbf{v}, \mathbf{v}', \mathbf{w}', \mathbf{w})$ is a parallelogram as well.

Proof: By Theorem 5, $\mathbf{v} = \mathbf{v}' - \mathbf{u}' + \mathbf{u}$, and $\mathbf{w} = \mathbf{w}' - \mathbf{u}' + \mathbf{u}$. (See Fig. 4.1). Thus $\mathbf{w} = \mathbf{w}' + (\mathbf{v} - \mathbf{v}') = \mathbf{w}' - \mathbf{v} + \mathbf{v}'$, and $(\mathbf{v}, \mathbf{v}', \mathbf{w}', \mathbf{w})$ is a parallelogram. ∎

■ **THEOREM 9. Desargues' Theorem II.** Suppose that $\ell(\mathbf{u}, \mathbf{u}') \cap \ell(\mathbf{v}, \mathbf{v}') \cap \ell(\mathbf{w}, \mathbf{w}') = \mathbf{z}$. Suppose further that $\ell(\mathbf{u}, \mathbf{v}) \| \ell(\mathbf{u}', \mathbf{v}')$ and $\ell(\mathbf{u}, \mathbf{w}) \| \ell(\mathbf{u}', \mathbf{w}')$. Then $\ell(\mathbf{v}, \mathbf{w}) \| \ell(\mathbf{v}', \mathbf{w}')$.

Proof: By Theorem 6, $\mathbf{u}' = \mathbf{u} + t(\mathbf{z} - \mathbf{u})$. If $\mathbf{x} = \mathbf{v} + t(\mathbf{z} - \mathbf{v})$, then $\mathbf{x}$ is on $\ell(\mathbf{v}, \mathbf{z})$ and $\ell(\mathbf{u}, \mathbf{v}) \| \ell(\mathbf{u}', \mathbf{x})$ by Theorem 7. Thus $\mathbf{x} = \mathbf{v}'$. Similarly $\mathbf{w}' = \mathbf{w} + t(\mathbf{z} - \mathbf{w})$. But this implies that $\mathbf{w}' - \mathbf{v}' = \mathbf{w} - \mathbf{v} + t(\mathbf{z} - \mathbf{w} - \mathbf{z} + \mathbf{v}) = (1 - t)(\mathbf{w} - \mathbf{v})$, so $\ell(\mathbf{v}, \mathbf{w}) \| \ell(\mathbf{v}', \mathbf{w}')$. (See Fig. 4.2). ∎

If $\mathbb{D}$ is a division ring, then the affine plane $(\mathscr{P}, \mathscr{L})$ described in Theorem 4 is called the *affine plane over* $\mathbb{D}$. Theorems 8 and 9 imply the following:

■ **THEOREM 10.** For any division ring $\mathbb{D}$, the affine plane over $\mathbb{D}$ is Desarguesian.

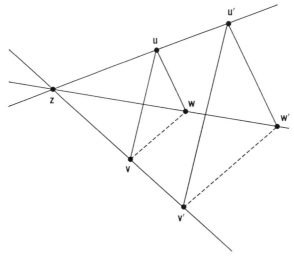

Figure 4.2

Proof: Theorems 9 and 10.   ∎

In later chapters the focus will be on higher-dimensional affine geometry. Just as three-dimensional or *solid* Euclidean geometry is really the
study of $\mathbb{R}^3$, *n*-dimensional affine geometry is the study of the vector
spaces $\mathbb{D}^n$. First, however, the affine planes of orders 4 and 5 are
reintroduced as $(GF(4))^2$ and $(\mathbb{Z}_5)^2$, respectively, where it should be noted
that the lines in each example satisfy *linear* equations with coefficients in
the appropriate field.

□**EXAMPLE 2: The affine plane** $(GF(4))^2$**.** The points here are the
ordered pairs from $GF(4)$ and the lines are the sets of points which satisfy
linear equations. Thus, since $GF(4) = \{0, 1, u, u + 1\}$ and $u^2 = u + 1$,
there are sixteen points in the plane $(GF(4))^2$. (Note that as in Appendix
A, we are using the variable u in place of x, since x and y are
traditionally used as the variables in linear equations.) Its twenty lines
comprise five pencils of parallel lines whose equations are given by

$$y = 0, 1, u, u + 1,$$
$$x = 0, 1, u, u + 1,$$
$$x + y = 0, 1, u, u + 1,$$
$$x + uy = 0, 1, u, u + 1,$$
$$x + (u + 1)y = 0, 1, u, u + 1.$$

For example, the line whose equation is $x + uy = u$ contains the points $(0, 1), (1, u), (u, 0)$, and $(u + 1, u + 1)$.   □

□**EXAMPLE 3: The affine plane $(Z_5)^2$.** Here the points are the twenty-five points in $(\mathbb{Z}_5)^2$ and equations for the lines are the linear equations with coefficients in $\mathbb{Z}_5$. The same line may satisfy more than one equation; for example, the line whose equation is $x + 2y = 1$ has exactly the same points as its multiple $3x + y = 3$. Thus it is easy to see that the equations are of the form $x + yb = c$ or $y = c$, for $b = 0, 1, 2, 3, 4$, and $c = 0, 1, 2, 3, 4$. As b and c take on different values, 30 lines result. Two lines are parallel exactly when they have equations of form $x + yb = c$ and $x + yb = d$, or $y = c$ and $y = d$ $(c \neq d)$, and each pencil of parallel lines has five lines.   □

Although it may not be obvious, a careful proof of Theorem 4 uses *all* of the axioms for division rings. For example, $(\mathbb{Z}_4)^2$ is not an affine plane. Its "points" $(0, 0)$ and $(2, 0)$ satisfy the equations $x = 0$ and $2x = 0$. But the "line" given by $x = 0$ has the four "points" $(0, 0), (0, 1), (0, 2)$, and $(0, 3)$, while these "points" and $(2, 0), (2, 1), (2, 2)$, and $(2, 3)$ satisfy $2x = 0$ as well. Thus the "lines" given by $x = 0$ and $2x = 0$ are different, and the "points" $(0, 0)$ and $(2, 0)$ lie on more than one "line"; so Axiom $A1$ for affine planes fails.

EXERCISES

1. Prove that, for any division ring $\mathbb{D}$ and positive integer n, addition is associative and commutative, $\mathbf{0}$ is the additive identity, and $(-x_1, \ldots, -x_n) = -1(x_1, \ldots, x_n)$ is the additive inverse for $(x_1, \ldots, x_n)$.
2. Prove that the following hold for $\mathbf{x}$ and $\mathbf{y}$ in $\mathbb{D}^n$ and a and b in $\mathbb{D}$:
   a. $a(\mathbf{x} + \mathbf{y}) = a\mathbf{x} + a\mathbf{y}$.
   b. $(a + b)\mathbf{x} + a\mathbf{x} + b\mathbf{x}$.
   c. $a(b\mathbf{x}) = (ab)\mathbf{x}$.
   d. $1\mathbf{x} = \mathbf{x}$ (where 1 is the multiplicative identity of $\mathbb{D}$).

3. Show that if two lines in $\mathbb{D}^2$ have two points in common, they are equal lines.
4. Show that for $a \neq 0$ in $\mathbb{D}$, the equations $x + yb = c$ and $xa + yba = ca$ determine the same line.
5. Show that two linear equations determine the same line if and only if they are *proportional*, in that each is a nonzero right multiple of the other. [*Hint:* Assume that $a \neq 0$, and multiply through by a^{-1} on the right. Now the equations are of the form $x + yb = c$ and $x + yd = e$. Show by solving the equations simultaneously that if $b \neq d$, there is a unique pair (x, y) satisfying both of them. Show that if $b = d$ and $c \neq e$, the equations have no points in common.]
6. Show that the line in $\mathbb{D}^2$ containing points $\mathbf{u}$ and $\mathbf{v}$ can be represented parametrically as
$$\{\mathbf{u} + t(\mathbf{v} - \mathbf{u}): t \in \mathbb{D}\}.$$

7. Suppose that $1 + 1 \neq 0$ in $\mathbb{D}$ and that (a, b), (c, d), (e, f), and $(a - c + e, b - d + f)$ are the vertices of a parallelogram. Prove that the diagonal containing (a, b) and (e, f) intersects the diagonal containing (c, d) and $(a - c + e, b - d + f)$ at the "midpoint" of each, that is, at the point $2^{-1}[(a, b) + (e, f)] = 2^{-1}[(c, d) + (a - c + e, b - d + f)]$.

8. List the points and equations for the lines in the affine plane over $\mathbb{Z}_7$. Find an equation for the line containing the points $(2, 6)$ and $(3, 2)$. Find an equation for the line through $(3, 1)$ parallel to it.

9. Show that $(\mathbb{Z})^2$ is not an affine plane by showing that one of the axioms for affine planes fails.

10. Recall that $GF(8) = \mathbb{Z}_2[u]/(u^3 + u + 1)$.

    a. Find an equation for the line through (u^2, u) and $(u + 1, u^2)$.

    b. Find an equation for the line through (u^2, u^2) parallel to the line in (a).

    c. Find a fourth vertex of a parallelogram containing $(1, 0)$, (u, u^2) and $(u, u^2 + 1)$ as three of its vertices.

11. Let $(\mathscr{P}, \mathscr{L})$ be the affine plane over the 125-element field $\mathbb{Z}_5[u]/(u^3 + u + 1)$.

    a. Compute its number of points, lines, and pencils of parallel lines.

    b. Are the points $(u, u + 1)$, $(3u, 2)$, and $(1, 1)$ collinear?

    c. Find an equation for the line containing $(2, u)$ and parallel to the line whose equation is $2x + uy = 4$.

    d. Find the point (x, y) such that $((u, u^2), (1, u), (u + 1, 3), (x, y))$ is a parallelogram.

    e. Find the point of intersection of the lines whose equations are $ux + u^2y = 1$ and $x + 2y = u$.

12. The field $GF(49)$ is isomorphic to $\mathbb{Z}_7[u]/(u^2 + 4)$. Find an equation for the line containing $(3, 1)$ and $(u + 1, u)$. Find the line containing $(3, u)$ which is parallel to it.

13. If $1 + 1 \neq 0$ in $\mathbb{D}$, define the "midpoint" of a pair of points $\mathbf{x}$ and $\mathbf{y}$ in $\mathbb{D}^2$ to be $2^{-1}(\mathbf{x} + \mathbf{y})$. Define a triangle to be a set of three noncollinear points (called its vertices) in $\mathbb{D}^2$. Suppose that $1 + 1 + 1 \neq 0$ in $\mathbb{D}$. Show that the "medians" (a line between a vertex and the "midpoint" of its opposite side is a median) of any triangle intersect. Furthermore the point of intersection is $3^{-1}\mathbf{x} + (1 - 3^{-1})\mathbf{m}$ where $\mathbf{x}$ is any vertex and $\mathbf{m}$ is the "midpoint" of its opposite side.

4.4 COORDINATIZING POINTS

As a first step toward coordinatization, it can now be shown that any two lines in a Desarguesian affine plane, besides having the same cardinality, are in fact isomorphic as division rings. Since we will be concerned with selecting different points to be the additive and multiplicative identities on lines (i.e., different O's and I's) we will refer to the ring formed from line ℓ and O and I as identities as (ℓ, O, I) rather than $(\ell, +, \times)$.

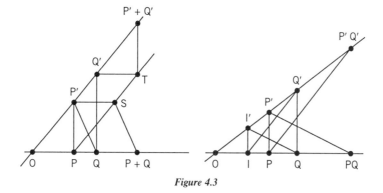

Figure 4.3

■ **THEOREM 11.** Let $\ell = \ell(O, I)$ and $m = \ell(O', I')$ be any two lines in a Desarguesian affine plane $(\mathscr{P}, \mathscr{L})$. Then $(\ell, O, I) \cong (m, O', I')$.

Proof:

CASE 1. Suppose that $\ell \cap m = O = O'$. We will show that $(\ell, O, I) \cong (m, O, I')$. (See Fig. 4.3.) Define $\phi: \ell \to m$ by $\phi(O) = O$, $\phi(I) = I'$, and for X different from O and I, $\phi(X) = X'$ where X' is the point of intersection of m with the line through X parallel to $\ell(I, I')$. Thus $\ell(X, X') \| \ell(Y, Y')$ whenever X and Y are distinct nonzero points on ℓ. It was shown in Theorem 2 of Chapter 2 that ϕ is one-to-one and onto.

To show that ϕ preserves addition, let P and Q be distinct points on ℓ. Compute P + Q using P' (and S). Compute Q' + P' using P (and T). The additions give $\ell(P', Q) \| \ell(S, P + Q)$ and $\ell(P, P') \| \ell(T, Q' + P')$. Use Desargues' Theorem on T, S, P + Q and Q', P', Q to obtain $\ell(T, P + Q) \| \ell(Q', Q)$. But $\ell(T, Q' + P') \| \ell(P, P') \| \ell(Q, Q')$. This implies that P + Q, T, and $Q' + P'(= P' + Q')$ are collinear and on a line parallel to $\ell(P, P')$. But $\ell(P + Q, (P + Q)') \| \ell(P, P')$. Thus $(P + Q)' = P' + Q'$, and ϕ preserves addition.

To show that ϕ preserves multiplication, compute PQ using I' (and P'), and compute P'Q' using I (and P). Now compute P'Q' using Q (and PQ). Then $\ell(Q, Q') \| \ell(PQ, P'Q')$, and it follows that $P'Q' = (PQ)'$. (The reader should check that addition and multiplication are preserved when P = Q as well.)

CASE 2. $\ell = m$, $O = O'$ I $\neq$ I'. Let $k \neq \ell$ by any line containing O, and let I* be a point of k different from O. By Case 1, $(\ell, O, I) \cong (k, O, I^*)$ and $(k, O, I^*) \cong (\ell, O, I')$. Thus $(\ell, O, I) \cong (\ell, O, I')$.

CASE 3. $\ell \| m$. Choosing, if necessary, a different point I', let $X = \ell(O, O') \cap \ell(I, I')$. Define $\phi(P) = P' = \ell(P, X) \cap m$ for any point P in ℓ. For P and Q distinct points on ℓ, add P + Q using X (and T) and add P' + Q' using X (and T'). Use the converse of Desargues' Theorem on

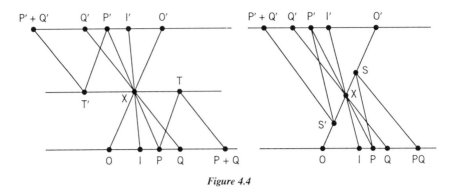

Figure 4.4

PT(P + Q) and P'T'(P' + Q') to get P' + Q', X and P + Q collinear, from which it follows that P' + Q' = (P + Q)' and ϕ preserves addition. To show the preservation of multiplication, multiply PQ using X (and S); multiply P'Q' using X (and S'). Use the converse of Desargues' Theorem on PS(PQ) and P'S' (P'Q') to get PQ, X and P'Q' collinear, which completes the proof of Case 3. (See Fig. 4.4.)

CASE 4. $\ell \cap m = X$ where X is different from O or O'. By Case 2, it can be assumed that X is different from I or I'. Let $\mathscr{k}$ be the line containing O' and parallel to ℓ. Let I* ≠ O' be a point on $\mathscr{k}$. Now $(\ell, O, I) \cong (\mathscr{k}, O', I^*) \cong (m, O', I')$ which completes the proof of the theorem. (See Fig. 4.5.) ∎

We now come to the main results of this chapter, namely the coordinatization of Desarguesian affine planes. The aim here is to identify each point P with an ordered pair of elements from a division ring, and each line ℓ with a linear equation over the same ring.

Given any Desarguesian affine plane $(\mathscr{P}, \mathscr{L})$, select and fix two distinct nonparallel lines, and call them ℓ_x and ℓ_y. Let $\ell_x \cap \ell_y$ be

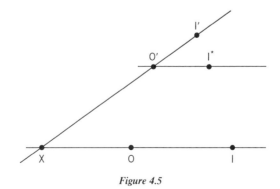

Figure 4.5

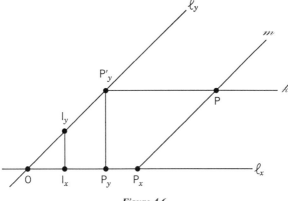

Figure 4.6

denoted O, and pick points I_x and I_y on ℓ_x and ℓ_y, respectively, each of which is different from O. By Theorem 11 (Case 1), $(\ell_x, O, I_x) \cong (\ell_y, O, I_y)$. Further any A′ different from O and I_y in ℓ_y is the image of the point A in ℓ_x such that $\ell(A, A') \| \ell(I_x, I_y)$.

Let P be a point in $\mathcal{P}$. Let m be the line containing P parallel or equal to ℓ_y, and let $P_x = m \cap \ell_x$. Let k be the line containing P and parallel or equal to ℓ_x. Let $P'_x = k \cap \ell_y$. Note that P_x and P_y are elements of the division ring ℓ_x. Rename P by the ordered pair (P_x, P_y) of elements from the division ring ℓ_x. (See Fig. 4.6.) Observe that for $P \in \ell_x$, the line m contains P, whence $P_x = P$, while k is the line ℓ_x, so $k \cap \ell_y = O$. Thus $P = (P, O)$. Similarly for $Q' \in \ell_y$, $Q' = (O, Q)$. For consistency I_x and I_y will be called I and I′, respectively. The lines ℓ_x and ℓ_y are called the x-axis and the y-axis.

The preceding results can be summarized in the following theorem.

■ **THEOREM 12.** For any Desarguesian affine plane $(\mathcal{P}, \mathcal{L})$ and axes $\ell_x = \ell(O, I)$ and $\ell_y = (O, I')$ in $\mathcal{L}$, the points of $\mathcal{P}$ are in one-one correspondence with the ordered pairs in $\ell_x \times \ell_x$.

EXERCISES

1. Complete the proof of Case 1 of Theorem 11 by showing that $\phi(P + P) = \phi(P) + \phi(P)$ and $\phi(P \cdot P) = \phi(P) \cdot \phi(P)$ for all $P \in \ell$.

2. Make addition and multiplication tables for the line {OIDF} in the affine plane of order 4 given in Example 8 of Chapter 2. Do the same for line {BEOL} with O again the additive identity and L = I′ the multiplicative identity. Find D′ and F′ on {BEOL}. Add D′ + F′ on {BEOL}, and show that the sum obtained as (D + F)′. Show by comparing the tables that the lines {OIDF} and {BEOL} are isomorphic as rings. Show that each is isomorphic to $\mathbb{Z}_2[u]/(u^2 + u + 1)$.

   Exercises 3–7 refer to the plane of order 5, first described in Chapter 3 and repeated here for convenience.

The affine plane of order 5. $\mathscr{P} = \{A, B, C, \ldots, Y\}$, and the lines are

{OIABC}	{OHMRV}
{DEFGH}	{IDNSW}
{JKLMN}	{AEJTZ}
{PQRST}	{BFKPY}
{UVWXY}	{CGLQU}
{ODJPU}	{OGKTW}
{IEKQV}	{IHLPX}
{AFLRW}	{ADMQY}
{BGMSX}	{BENRU}
{CHNTY}	{CFJSV}
{OELSY}	{OFNQX}
{CDKRX}	{IGJRY}
{BHJQW}	{AHKSU}
{AGNPV}	{BDLTV}
{IFMTU}	{CEMPW}

3. Show an explicit isomorphism between the rings {OIABC} and {OHMRV} with I′ = H.

4. Show an explicit isomorphism between the rings {OIABC} and {JKLMN} with O′ = K and I′ = J.

5. Show an explicit isomorphism between the rings {OIABC} and {ADMQY} with O′ = Q and I′ = Y.

6. In the affine plane of order 5 given above, let ℓ_x = {OIABC} and ℓ_y = {ODJPU} with I′ = J. Coordinatize the point F (i.e., rename it as an ordered pair from {OIABC}). Which point has coordinates (C, A)? Which has coordinates (O, B)?

7. Find an ordered pair (besides (O, O)) of elements from {OIABC} that satisfies the equation $x + yC = O$. Find the element in $\mathscr{P}$ that corresponds to this pair. Consider the line through this point and (O, O). Show that all five points on this line satisfy $x + yC = O$.

8. Let $(\mathscr{P}, \mathscr{L})$ be the nine-point, twelve-line plane given in Example 7 of Chapter 2 whose points are $\{A, B, \ldots, I\}$. Let O = B, I = I, and I′ = C. Find ℓ_x, ℓ_y, and coordinatize the nine points of the plane.

4.5 LINEAR EQUATIONS

From the preceding coordinatization of points, it is clear that the points O, I, and I′ are coordinatized as (O, O), (I, O), and (O, I), respectively. Any point on ℓ_x is of the form (P, O); thus every point on ℓ_x satisfies the equation $y = O$. Similarly ℓ_y has equation $x = O$. If a line is parallel to ℓ_x, its points all have the same second coordinate; therefore the line satisfies the equation $y = Q$ for some Q in the ring ℓ_x. Similarly line parallel to ℓ_y have equations of the form $x = P$.

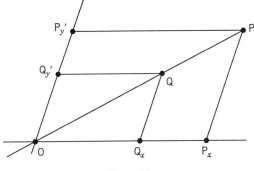

Figure 4.7

To show that the points on each line satisfy a linear equation with coefficients in ℓ_x, the following preliminary geometric result is needed.

■ **THEOREM 13.** Let $(\mathcal{P}, \mathcal{L})$ be a Desarguesian affine plane, with ℓ_x and ℓ_y given as above. Let m be a line containing O, but different from ℓ_x and ℓ_y. Then for any pair of distinct points $P = (P_x, P_y)$ and $Q = (Q_x, Q_y)$ in m, $\ell(P'_y, P_x) \| \ell(Q'_y, Q_x)$.

Proof: Use Desargues' Theorem on $PP'_y P_x$ and $QQ'_y Q_x$. (See Fig. 4.7.) ■

Because of Theorem 13, to any line m containing O but different from ℓ_x and ℓ_y can be associated a point M in ℓ_x where $\ell(I', M)$ is parallel or equal to $\ell(P'_y, P_x)$ for any nonzero (P_x, P_y) in m, as shown in Fig. 4.8.

■ **THEOREM 14.** Let m be a line containing O but different from ℓ_x and ℓ_y, with M the point in ℓ_x such that $\ell(I', M) \| \ell(P'_y, P_x)$ for every $P = (P_x, P_y)$ in m. Then every point on m satisfies the equation $x = yM$.

Proof: Let $P = (P_x, P_y) \in m$ be different from O. Multiply $P_y M$ using I' and P'_y as auxiliary points, as shown in Fig. 4.9. Since $\ell(P'_y, P_x)$ is parallel

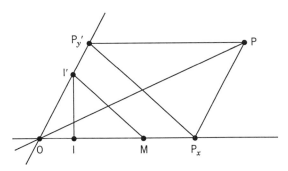

Figure 4.8

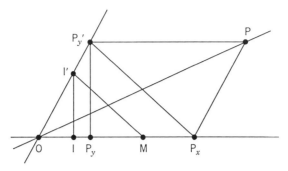

Figure 4.9

or equal to $\angle(I', M)$, it follows that $P_x = P_y M$. Since $O = OM$ for any M, $O = (O, O)$ also satisfies $P_x = P_y M$; hence every point on m satisfies the equation. ∎

■ **THEOREM 15.** If m and M are defined as in Theorem 14, then for any (X, Y) in $\mathscr{P}$ with $X = YM$, $(X, Y) \in m$.

Proof: Let (P_x, P_y) be a point in m different from O. If $(P_x, P_y) = (X, Y)$, then (X, Y) is in m, and the proof is complete. Otherwise, multiplying YM using I' and P_y' gives $\angle(I', M)$ parallel or equal to $\angle(Y', X)$; thus $\angle(I', M)$ is parallel to $\angle(P_y', P_x)$. Now use the converse of Desargues' Theorem on $Y'(X, Y)X$ and $P_y' P P_x$ to get $\angle(Y', P_y')$, $\angle((X, Y), P)$ and $\angle(X, P_x)$ either parallel or meeting at one point. Since the first and third lines meet at O, $\angle((X, Y), P)$ contains O. Therefore it is equal to the line containing P and O, which is m. (See Fig. 4.10.)

■ **THEOREM 16.** Suppose that k is a line in $\mathscr{L}$ not containing O, not parallel in ℓ_y, but intersecting ℓ_x at point C. Let m be the line

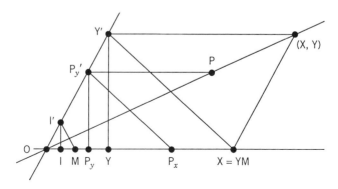

Figure 4.10

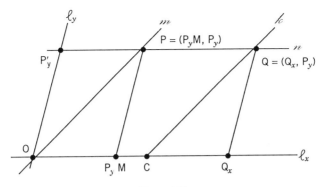

Figure 4.11

through O parallel to k. Then, if m satisfies the equation $x = yM$, k satisfies the equation $x = yM + C$.

Proof: Suppose that Q (different from C), is in k, and let n be the line through Q parallel to ℓ_x. Suppose that $n \cap m = P = (P_yM, P_y)$. Then Q has coordinates (Q_x, P_y), as shown in Fig. 4.11. Add C + P_yM using P and Q. Since $\angle(P, P_yM) \| \ell_y \| \angle(Q, Q_x)$, we have $Q_x = C + P_y$M. Thus $Q = (Q_x, P_y)$ satisfies the equation $x = C + yM = yM + C$, since addition is commutative. We leave as an exercise the proof that any point (X, Y) whose coordinates satisfy the equation $x = yM + C$ is on k.   ■

The results on lines can be summarized in the following theorem:

■ **THEOREM 17.** Suppose that $(\mathscr{P}, \mathscr{L})$ is a Desarguesian affine plane, with ℓ_x and ℓ_y selected as axes. Then

    i. ℓ_x has equation $y = O$,
    ii. ℓ_y has equation $x = O$,
    iii. If $O \in m$, m has equation $x = yM$,
    iv. If $k \| m$ and $k \cap \ell_x = C$, k has equation $x = yM + C$.

EXERCISES

1. Let $(\mathscr{P}, \mathscr{L})$ be the sixteen-point, twenty-line plane with $\ell_x = \{OIDF\}$, $\ell_y = \{BEOL\}$, and $I' = L$. Which point on ℓ_x is playing the role of M for line $\{CKGO\}$? Find an equation for $\{CKGO\}$. Find an equation for $\{DJLP\}$.
Exercises 2–9 refer to the plane of order 5 given before Exercise 3 of Section 4.4, p. 86.

2. Let $\ell_x = \{OIABC\}$ and $\ell_y = \{ODJPU\}$ with $I' = J$. Show that ℓ_x is isomorphic to $\mathbb{Z}_5$. Rename each point of ℓ_x as an element of $\mathbb{Z}_5$ (O will be 0, I will be 1, I + I will be 2, etc.).

3. Rename each point of ℓ_y as the prime of an element of $\mathbb{Z}_5$, recalling that $\ell(1, 1') \| \ell(2, 2')$, and so on.

4. Coordinatize the points C, U, K, Q, and T of the plane as ordered pairs of elements from $\mathbb{Z}_5$. [*Example*: The line through K parallel to ℓ_y is {IEKQV} which meets ℓ_x at I = 1. The first coordinate of K is 1. The line through K parallel to ℓ_x is {JKLMN} which meets ℓ_y at J = I' = 1'. Thus the second coordinate of K is 1 as well, and K = (1, 1).]

5. What is an equation for {OGKTW}? Recall that this line contains O = (0, 0) and K = (1, 1). Since each pair of points is on a unique line, finding any linear equation satisfied by both O and K will suffice.

6. What is an equation for {ADMQY}? Note that this line is parallel to {OGKTW} whose equation found in Exercise 5.

7. Find equations for the lines in the pencil containing {OELSY}.

8. Which point has coordinates (3, 2)?

9. Which line has equation $x = 3y + 2$?

4.6 THE THEOREM OF PAPPUS

Recall that if Pappus' Theorem holds in affine plane, then multiplication of coordinates on any line of that plane is commutative. The converse of this result can now be proved, based on coordinatization.

■   **THEOREM 18.** Let $(\mathscr{P}, \mathscr{L})$ be the affine plane over the division ring $\mathbb{D}$. Then Pappus' Theorem is true in $(\mathscr{P}, \mathscr{L})$ if and only if $\mathbb{D}$ is a field.

Proof: It need only be shown that Pappus' Theorem holds when multiplication is commutative on $\mathbb{D}$. Suppose that $\ell \cap m = X$, then P, Q, R in ℓ and S, T, U in m. Assume $\ell(P, T) \| \ell(Q, U)$ and $\ell(Q, S) \| \ell(R, T)$. It must be shown that $\ell(P, S) \| \ell(R, U)$. Since all lines are isomorphic as division rings, I can be chosen on ℓ such that $\ell(I, S) \| \ell(P, T)$, and X can be selected on O. Now $(\ell, O, I) \cong \mathbb{D}$, whence multiplication is commutative on ℓ. But PQ (multiplied using S (and T)) is R; thus QP is R. Therefore $\ell(P, S) \| \ell(R, U)$. (See Fig. 4.12.)   ■

It is now possible to construct an example of an affine plane in which Pappus' Theorem fails. By Wedderburn's Theorem, every *finite* division ring is a field; thus Pappus' Theorem is true in all finite Desarguesian affine planes. However, recall the noncommutative division ring $\mathbb{H}$ of quaternions introduced in Example 1. These can be represented as the elements $a + bi + cj + dk$, where $i^2 = j^2 = k^2 = -1$ and $ij = -ji = k$, $jk = -kj = i$, and $ki = -ik = j$.

Let $(\mathscr{P}, \mathscr{L})$ be the affine plane over the quaternions. It is easy to find a pair of quaternions that do not commute; for example, $ij \neq ji$. Let P = $(i, 0)$, and Q = $(j, 0)$, O = $(0, 0)$, I = $(1, 0)$, S = $(0, 1)$, T = $(0, i)$ U =

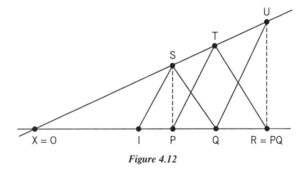

Figure 4.12

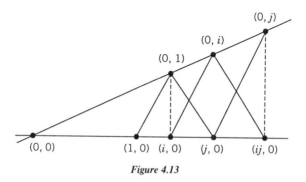

Figure 4.13

$(0, j)$, and $R = (ij, 0)$. Then in the affine plane over the $\mathbb{H}$, the dotted lines in Fig. 4.13 cannot be parallel; otherwise, R would be $(ji, 0)$, a contradiction.

Recall that the characteristic of a field is the smallest integer n such that $n \cdot 1 = 1 + \cdots + 1$ (n times) $= 0$. If no such n exists, the field is said to have characteristic 0. The *characteristic of a division ring* can be defined in exactly the same way and, like the characteristic of a field, must be 0 or a prime number.

■ **THEOREM 19.** The characteristic of a division ring is either zero or a prime number.

Proof: First note that, in a division ring, if $ab = 0$, then a or b is zero. This is because if $a \neq 0$, then $a^{-1}(ab) = a^{-1} \cdot 0 = 0$, while $(a^{-1}a) \cdot b = b$; thus $b = 0$. Now the proof follows from the fact that $(p1)(q1) = (pq)1$, which results from distributivity.   ■

There is a nice geometric characterization of affine planes whose coordinatizing division rings have characteristic 2. Recall that a parallelogram in an affine plane is an ordered quadruple of points (A, B, C, D) such

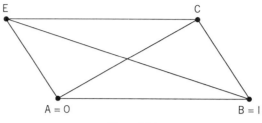

Figure 4.14

that $\ell(A, B) \| \ell(C, D)$ and $\ell(A, D) \| \ell(B, C)$. The *diagonals* of the parallelogram (A, B, C, D) are the lines $\ell(A, C)$ and $\ell(B, D)$.

■   **THEOREM 20.** Let $(\mathscr{P}, \mathscr{L})$ be the affine plane over the division ring $\mathbb{D}$. The following three statements are equivalent:

    i. The diagonals of some parallelogram in $(\mathscr{P}, \mathscr{L})$ are parallel.

    ii. The characteristic of $\mathbb{D}$ is 2.

    iii. The diagonals of every parallelogram in $(\mathscr{P}, \mathscr{L})$ are parallel.

Proof: i → ii: Suppose that the diagonals of (A, B, C, E) are parallel, as shown in Fig. 4.14.. Let $A = O$, $B = I$, and add $I + I$ using E (and C). Since $(E, I) \| \ell(C, O)$, then $I + I = O$, and the characteristic of $\mathbb{D}$ is 2.

    ii → iii: Let the characteristic of $\mathbb{D}$ be 2, and let (A, B, C, E) be a parallelogram in $(\mathscr{P}, \mathscr{L})$. Set $A = O$, $B = I$, and add $I + I$ using C (and E). Since $I + I = O$, $\ell(E, I)$ and $\ell(C, O)$ are parallel, and these are the diagonals of (A, B, C, E).

    iii → i: The proof is trivial.   ■

EXERCISES

1. Prove that the characteristic of the field $GF(p^n)$ is p.

2. Show that for a, b, c, and m elements of a division ring

    a. $\ell((0, 0), (a, m)) \| \ell((0, b), (ca, cm + b))$.

    b. $((0, b), (b, b), (a + b, a + b), (a, a + b))$ is a parallelogram whenever $0 \neq b$.

    c. Show that $((0, 0), (a, a), (a, a + b), (0, b))$ is a parallelogram whenever $0 \neq a, b$.

3. Find an example of the Pappus configuration in the synthetic affine plane of order 5. Verify that Pappus' Theorem holds for this configuration.

SUGGESTED READING

G. Birkhoff and S. Mac Lane, *A Survey of Modern Algebra*, 4th ed., New York: Macmillan, 1977.

D. Hilbert, *Foundations of Geometry*, 10th ed., La Salle, IL: Open Court, 1971.

CHAPTER 5

Coordinate Projective Planes

In Example 10 of Chapter 2, the *real projective plane* was defined as having equivalence classes of three-dimensional real vectors as its points, and lines satisfying homogeneous linear equations. In section 5.3 we will see that any Desarguesian projective plane can be represented in this way, but arbitrary division rings will be used instead of the real numbers.

5.1 PROJECTIVE POINTS AND HOMOGENEOUS EQUATIONS IN $\mathbb{D}^3$

Projective points can be defined to be equivalence classes of vectors in three-space, under the following equivalence relation.

Definition. For $\mathbb{D}$ a division ring, define $\sim$ on the nonzero vectors of $\mathbb{D}^3$ by $(p, q, r) \sim (s, t, u)$ if and only if $(p, q, r) = m(s, t, u)$ for some nonzero $m \in \mathbb{D}$.

Clearly $\sim$ is an equivalence relation; $(p, q, r) = 1(p, q, r)$ so it is reflexive; if $(p, q, r) = m(s, t, u)$, then $(s, t, u) = m^{-1}(p, q, r)$, so it is symmetric; and if $(p, q, r) = m(s, t, u)$ and $(s, t, u) = k(v, w, z)$, then $(p, q, r) = mk(v, w, z)$, so it is transitive.

Definition. The *projective point* $\langle p, q, r \rangle$ is defined to be the equivalence class containing the vector (p, q, r) in $\mathbb{D}^3$.

An equation of the form $xa + yb + zc = 0$, with $a, b, c \in \mathbb{D}$, and not all zero is called a *homogeneous linear equation*. It is *homogeneous* since

93

each of the variables x, y, z has the same exponent, and *linear* because the exponent is 1.

As was the case for the affine linear equations, the coefficients here are written on the right, so that the following theorem holds.

■ THEOREM 1

   i. If (p, q, r) satisfies $xa + yb + zc = 0$, then every vector equivalent to it under $\sim$ satisfies $xa + yb + zc = 0$.
   ii. The vector (p, q, r) satisfies $xa + yb + zc = 0$ if and only if it also satisfies $xam + ybm + zcm = 0$ for every nonzero $m \in \mathbb{D}$.

Proof: Since $pa + qb + rc = 0$, then $mpa + mqb + mrc = m \cdot 0 = 0$ for any $m \in \mathbb{D}$, so part i holds. Furthermore, $pam + qbm + rcm = 0 \cdot m = 0$; conversely, if $pam + qbm + rcm = 0$ and $m \neq 0$, then $pa + qb + rc = 0 \cdot m^{-1} = 0$, so part ii obtains.   ■

If $\mathbb{D}$ is a field, the equations can be written in the more familiar form $ax + by + cz = 0$. In view of Theorem 1, we make the following definition.

Definition

   1. The projective point $\langle p, q, r \rangle$ is said to *satisfy* the homogeneous linear equation $xa + yb + zc = 0$ when any one (and therefore all) of the vectors in $\langle p, q, r \rangle$ satisfies the equation.
   2. Two homogeneous linear equations are said to be *equivalent* if they are of the form $xa + yb + zc = 0$ and $xam + ybm + zcm = 0$ for some nonzero $m \in \mathbb{D}$.

The equivalence defined for homogeneous linear equations in part 2 of the definition above is an equivalence relation.

The following corollary is immediate from Theorem 1.

Corollary. The projective point $\langle p, q, r \rangle$ satisfies $xa + yb + zc = 0$ if and only if it satisfies every equivalent homogeneous linear equation.

Duality

Clearly every projective point is determined by a nonzero vector (p, q, r) and contains all the nonzero left multiples of (p, q, r). Similarly every homogeneous linear equation $xa + yb + zc = 0$ is determined by the nonzero vector (a, b, c) and is equivalent to all the equations determined by the nonzero *right* multiples (am, bm, cm). Thus there is a *duality* between projective points and equivalent homogeneous equations that reverses left and right scalar multiplication.

■ **THEOREM 2**

i. The projective points in $\mathbb{D}^3$ can be described as $\{\langle p, q, 1 \rangle : p, q, \in \mathbb{D}\} \cup \{\langle p, 1, 0 \rangle : p \in \mathbb{D}\} \cup \{\langle 1, 0, 0 \rangle\}$.

ii. Every homogeneous linear equation is equivalent to a unique one of the form $x + yb + cz = 0$, $y + zc = 0$, or $z = 0$.

Proof: (i) For any equivalence class $\langle p, q, r \rangle$, if $r \neq 0$, then $\langle p, q, r \rangle = \langle r^{-1}p, r^{-1}q, 1 \rangle$; if $r = 0$, and $q \neq 0$, it is equal to $\langle q^{-1}p, 1, 0 \rangle$. If q and r are both zero, then $p \neq 0$, so $\langle p, q, r \rangle = \langle p, 0, 0 \rangle = \langle 1, 0, 0 \rangle$.

(ii) The equation $xa + yb + zc = 0$ is equivalent to $x + yba^{-1} + zca^{-1} = 0$ if $a \neq 0$, to $y + zcb^{-1} = 0$ if $a = 0$ and $b \neq 0$, and to $z = 0$ if $a = b = 0$.   ■

Definition. An equivalence class described as $\langle p, q, 1 \rangle$, $\langle p, 1, 0 \rangle$, or $\langle 1, 0, 0 \rangle$ is said to be in *canonical form*. Dually, a homogeneous linear equation of the form $x + yb + zc = 0$, $y + zc = 0$, or $z = 0$ is said to be in *canonical form*.

Subspaces of $\mathbb{D}^3$

Recall that a subspace of an arbitrary vector space $\mathscr{V}$ is a subset $\mathscr{S}$ of $\mathscr{V}$ which is closed under vector addition and scalar multiplication. The origin $\mathbf{0}$ is always a subspace and is contained in every subspace of $\mathscr{V}$. A one-dimensional subspace consists of a nonzero vector and all of its left scalar multiples; thus the projective points $\langle p, q, r \rangle$ correspond to the one-dimensional subspaces of $\mathbb{D}^3$ in that the one-dimensional subspaces are exactly the equivalence classes with $\mathbf{0}$ adjoined. Given a homogeneous linear equation $xa + yb + zc = 0$, its solution set of vectors is closed under left scalar multiplication. But if (p, q, r) and (s, t, u) are solutions, clearly $(p + s)a + (q + t)b + (r + u)c = 0$, so the solution set is a subspace of $\mathbb{D}^3$, namely a *two-dimensional* subspace. The next section will show that these one- and two-dimensional subspaces of $\mathbb{D}^3$ correspond to the points and lines of a projective plane.

EXERCISES

1. Prove that the equivalence defined for homogeneous linear equations is an equivalence relation.

2. In the following exercises let $\mathbb{D} = \mathbb{Z}_7$.

   a. Show that there are 57 projective points $\langle p, q, r \rangle$.

   b. List the projective points satisfying the homogeneous equation $x + 5y + z = 0$. (Note that since $\mathbb{Z}_7$ is commutative, the coefficients in the equations can be written on the left.)

   c. Find the unique point that satisfies $x + 5y + z = 0$ and $3x + 2y + z = 0$, and write it in canonical form.

    d. Find a homogeneous linear equation satisfied by $\langle 1, 5, 1 \rangle$ and $\langle 3, 2, 1 \rangle$, and write it in canonical form. Compare with (c).

3. In the following exercises, let $\mathbb{D} = GF(4) \cong \mathbb{Z}_2[u]/(u^2 + u + 1)$.

    a. List the projective points satisfying the homogeneous equation $ux + (u + 1)y + z = 0$.

    b. List the homogeneous equations satisfied by $\langle u, u + 1, 1 \rangle$. Compare with (a).

    c. Define the dual to a point $\langle p, q, r \rangle$ to be the equation $px + qy + rz = 0$. Which points satisfy their dual equations?

5.2 COORDINATE PROJECTIVE PLANES

Taking as points the equivalence classes $\langle p, q, r \rangle$ and lines as collections of points satisfying homogeneous linear equations gives a projective plane. Conversely, as will be shown in Section 5.3, every Desarguesian projective plane can be coordinatized in this way.

■ **THEOREM 3.** Given a division ring $\mathbb{D}$, $(\mathscr{P}', \mathscr{L}')$ is a projective plane, where $\mathscr{P}' = \{\langle p, q, r \rangle : (p, q, r) \in \mathbb{D}^3, (p, q, r) \neq (0, 0, 0)\}$ and $\ell' \in \mathscr{L}'$ if and only if ℓ' is the set of points $\langle p, q, r \rangle$ satisfying $xa + yb + zc = 0$.

    Proof: Axiom *P*1: Assume that $\langle p, q, r \rangle$ and $\langle s, t, u \rangle$ are in canonical form. To find coefficients a, b, c, for the equation $xa + yb + zc = 0$, we must solve the following simultaneous equations for a, b, and c.

$$pa + qb + rc = 0,$$

$$sa + tb + uc = 0.$$

    CASE 1. Assume that $r = u = 1$. Then subtracting gives $(q - t)b = (s - p)a$. If $q = t$, then $s \neq p$, and $a = 0$. Set $b = 1$, and solve $qb + c = 0$ for c. If $q \neq t$, then set $b = 1$ and solve $q - t = (s - p)a$ for a.

    CASE 2. Assume that $r = 1$ and $u = 0$. Then we have

$$pa + qb + c = 0,$$

$$sa + b = 0.$$

Set $a = 1$, and solve the second equation for b. Substitute in the first equation to get c.

    CASE 3. Assume that $r = u = 0$. Then we have

$$pa + b = 0,$$

$$sa + b = 0.$$

Since $p \neq s$, a must equal 0. Thus $b = 0$ as well. Therefore $c = 1$, and the equation $z = 0$ is satisfied by $\langle p, 1, 0 \rangle$ and $\langle s, 1, 0 \rangle$.

Uniqueness follows since choosing a different b in Case 1, a different a in Case 2, or a different c in Case 3 would give equivalent equations.

Axiom $P2$: Assume that $xa + yb + zc = 0$ and $xd + ye + zf = 0$ are in canonical form. To find their point of intersection, we must solve for x, y, and z.

$$xa + yb + zc = 0,$$

$$xd + ye + zf = 0.$$

The proof is analogous to that for Axiom $P1$, with the three cases (1) $a = d = 1$, (2) $a = 1$ and $d = 0$, and (3) $a = d = 0$.

Axiom $P3$: For the homogeneous equation $x + yb + zc = 0$, assuming that neither b nor c is zero, $\langle 0, c, - cbc^{-1} \rangle$, $\langle 1, 0, - c^{-1} \rangle$, and their sum $\langle 1, c, - cbc^{-1} - c^{-1} \rangle$ are three distinct solutions. For $y + zc = 0$, with $c \neq 0$, $\langle 1, 0, 0 \rangle$, $\langle 0, - c, 1 \rangle$, and $\langle 1, - c, 1 \rangle$ are distinct solutions, and the equation $z = 0$ has $\langle 1, 1, 0 \rangle$, $\langle 1, 0, 0 \rangle$, and $\langle 0, 1, 0 \rangle$ as distinct solutions. Thus equations with 3, 2, or 1 nonzero coefficients have at least three distinct solutions, so there are at least three points on each line. Since $x = 0$ and $y = 0$ are inequivalent equations, there are at least two lines. ∎

Notation. It will be convenient below to denote the line whose equation is $xa + yb + zc = 0$ as $[a, b, c]$.

Corollary. $(\mathscr{P}', \mathscr{L}')$ is a projective plane where

$$\mathscr{P}' = \{\langle p, q, 1 \rangle : p, q \in \mathbb{D}\} \cup \{\langle p, 1, 0 \rangle : p \in \mathbb{D}\} \cup \{\langle 1, 0, 0 \rangle\},$$

$$\mathscr{L}' = \{[1, b, c] : b, c \in \mathbb{D}\} \cup \{[0, 1, c] : c \in \mathbb{D}\} \cup \{[0, 0, 1]\},$$

and $\langle p, q, r \rangle$ is on $[a, b, c]$ if and only if $pa + qb + rc = 0$. (Note the connection between the containment of a point on a line and the dot product of the three-dimensional vectors (p, q, r) and (a, b, c).)

Definition. The projective plane described in Theorem 3 and its corollary is called the *projective plane over* $\mathbb{D}$. It is denoted $PG_2(\mathbb{D})$.

☐ **EXAMPLE 1: The projective plane over $\mathbb{Z}_2$.** This is the same as the Fano plane, first introduced as Example 5 of Chapter 2, and labeled as a coordinate plane in Fig. 5.1. Its lines have the following equations: $x + y + z = 0$, $x + y = 0$, $x + z = 0$, $x = 0$, $y + z = 0$, $y = 0$, $z = 0$. ☐

☐ **EXAMPLE 2: The projective plane over $\mathbb{Z}_3$.** The points are the nine classes of the form $\langle p, q, 1 \rangle$, the three classes of the form $\langle p, 1, 0 \rangle$, and the class $\langle 1, 0, 0 \rangle$ where p and q are in $\mathbb{Z}_3$. Its lines satisfy the equations

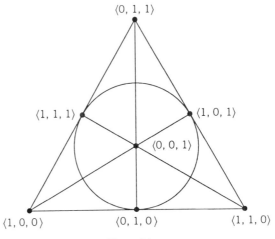

Figure 5.1

$x + by + cz = 0$ $(b, c \in \mathbb{Z}_3)$, $y + cz = 0$ $(c \in \mathbb{Z}_3)$, and $z = 0$. Note that since $\mathbb{Z}_3$ is a field, the equations may be written with the coefficients on the left. $\square$

EXERCISES

1. Show that the coordinate projective plane over the finite field $GF(p^k)$ has $p^{2k} + p^k + 1$ points and $p^{2k} + p^k + 1$ lines. How many points are on each line? How many lines pass through each point?

2. a. Let F be a field. Show that the equation for the line containing the distinct points $\langle a, b, c \rangle$ and $\langle d, e, f \rangle$ is

$$\begin{vmatrix} a & b & c \\ d & e & f \\ x & y & z \end{vmatrix} = 0$$

   in which the left side is a 3×3 determinant.

   b. Find a similar way of determining the point of intersection of two distinct lines in the projective plane over F.

   *c. Do analogous results hold for a noncommutative division ring $\mathbb{D}$?

3. The field $GF(49)$ is isomorphic to $\mathbb{Z}_7[u]/(u^2 + 4)$. The following exercises pertain to the projective plane over this field.

   a. Find an equation for the line containing $\langle u, 3u, 1 \rangle$ and $\langle 2u, u, 1 \rangle$.

   b. Find the point of intersection of the lines whose equations are

$$3ux + 6y + uz = 0$$

   and

$$(u + 2)x + 4uy + (u + 2)z = 0.$$

5.3 COORDINATIZATION OF DESARGUESIAN PROJECTIVE PLANES

The coordinatization of Desarguesian affine planes described in Chapter 4 can be extended to coordinatize Desarguesian projective planes. First, it can be shown formally that removal of a line at infinity from a Desarguesian projective plane results in a Desarguesian affine plane.

■ **THEOREM 4.** Suppose that $(\mathscr{P}', \mathscr{L}')$ is a Desarguesian projective plane. Select a line at infinity, ℓ_∞, and remove it to obtain the affine plane $(\mathscr{P}, \mathscr{L})$. Then $(\mathscr{P}, \mathscr{L})$ is Desarguesian.

Proof: By Theorems 9 and 10 of Chapter 3, it is necessary only to prove that Desargues' Theorem II holds in $(\mathscr{P}, \mathscr{L})$. Suppose that $\ell(A, A') \cap \ell(B, B') \cap \ell(C, C') = P$, $\ell(A, B) \| \ell(A', B')$, and $\ell(A, C) \| \ell(A', C')$, where A, B, and C are noncollinear points. Then in $(\mathscr{P}', \mathscr{L}')$, $\ell'(A, A') \cap \ell'(B, B') \cap \ell'(C, C') = P$, so the triangles ABC and A'B'C' are perspective from a point. Therefore they are perspective from a line m'. But, since $\ell'(A, B) \cap \ell'(A', B') \in \ell_\infty$ and $\ell'(A, C) \cap \ell'(A', C') \in \ell_\infty$, $m' = \ell_\infty$. But this implies that $\ell'(B, C) \cap \ell'(B', C') \in \ell_\infty$, so $\ell(B, C) \| \ell(B', C')$.

Coordinatization of Points

Given a Desarguesian projective plane $(\mathscr{P}', \mathscr{L}')$, select ℓ_∞, and remove it to obtain the Desarguesian affine plane $(\mathscr{P}, \mathscr{L})$. Now $\mathscr{P}' = \mathscr{P} \cup \ell_\infty$. The points of $\mathscr{P}$ can be represented as ordered pairs (x, y) from a division ring $\mathbb{D}$, while its lines satisfy linear equations of the form $xa + yb + c = 0$. Let $A \in \mathscr{P}'$. If A is in $\mathscr{P}$, $A = (x, y)$ for some x and y in $\mathbb{D}$. The *homogeneous coordinates* of A are defined to be $\langle x, y, 1 \rangle$. In particular, the origin $O \in \mathscr{P}$ has homogeneous coordinates $\langle 0, 0, 1 \rangle$, $I = \langle 1, 0, 1 \rangle$ and $I' = \langle 0, 1, 1 \rangle$. If $B \in \ell_\infty$, select any third point $C \in \ell'(O, B)$. Then $C \in \mathscr{P}$, so $C = (u, v)$ for some u and v in $\mathbb{D}$. Coordinatize B as $\langle u, v, 0 \rangle$.

■ **THEOREM 5.** The coordinatization of $B \in \ell_\infty$ does not depend on the choice of $C \in \ell'(O, B)$ (see Fig. 5.2).

Proof: Suppose that E is a fourth point on $\ell'(O, B)$. Then $E \in \mathscr{P}$, and as an affine point has coordinates (p, q). Since, as affine points O, E, and C are collinear and distinct and $O = (0, 0)$, E is a nonzero multiple of C, so $(p, q) = (mu, mv)$ for some nonzero m in $\mathbb{D}$. Thus $\langle p, q, 0 \rangle = \langle mu, mv, 0 \rangle = \langle u, v, 0 \rangle$ and the homogeneous coordinates of B are well defined.   ■

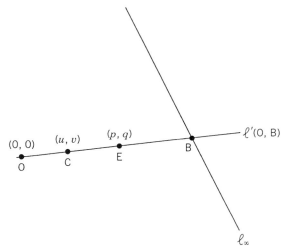

Figure 5.2

Coordinatization of Lines

In the coordinatization of the projective points above, each point in $\mathscr{P}$ has third coordinate 1, while each point in ℓ_∞ has third coordinate zero. Thus ℓ_∞ satisfies the homogeneous equation $z = 0$.

If $\ell' \neq \ell_\infty$, then ℓ, the line obtained from ℓ' by removing its point at infinity, is an affine line and satisfies an equation $xa + yb + c = 0$. Therefore for any point $A = (p, q) \in \ell$, it follows that $pa + qb + c = 0$; therefore $pa + qb + 1c = 0$, and $\langle p, q, 1 \rangle$ satisfies

$$xa + yb + zc = 0. \tag{5.1}$$

Suppose that $B = \ell' \cap \ell_\infty$. Then the affine line ℓ is parallel to the unique line m through the origin whose equation is $xa + yb = 0$. Thus $B \in m'$, so B has projective coordinates $\langle u, v, 0 \rangle$, where (u, v) is any affine point on m. Thus $ua + vb = 0$. Therefore $ua + vb + 0c = 0$ for any c, and $B = \langle u, v, 0 \rangle$ satisfies (5.1). Thus every point on ℓ' satisfies (5.1).

If B is in another line k', then since $k \| \ell$, k has the equation $xa + yb + d = 0$ for some $d \in \mathbb{D}$. Thus $0 = ua + vb = ua + vb + 0d$, and $\langle u, v, 0 \rangle$ satisfies $xa + yb + zd = 0$.

Suppose, conversely, that $xa + yb + zc = 0$ is any homogeneous linear equation over $\mathbb{D}$. Then, if $a = b = 0$, this is the equation for ℓ_∞; otherwise, $xa + yb + c = 0$ is the equation for an affine line ℓ in $\mathscr{L}$, thus $xa + yb + zc = 0$ is the equation for the line ℓ' in $\mathscr{L}'$.

These results can be summarized in the following theorem.

■ **THEOREM 6.** The points and lines of $(\mathscr{P}', \mathscr{L}')$ are in one-one correspondence with the points and lines of the coordinate projective plane over $\mathbb{D}$. Further $A \in \ell'$ if and only if the homogeneous coordinates $\langle p, q, r \rangle$ of A satisfy the homogeneous linear equation for ℓ'.

The following example presents the coordinatization of the projective plane of order 3 with its 13 ($= 3^2 + 3 + 1$) points and its 13 lines.

☐ **EXAMPLE 3: The thirteen-point, thirteen-line projective plane.** $\mathscr{P}' = \{A, B, C, D, E, F, G, H, J, K, L, M, N, P\}$, and $\mathscr{L}'$ consists of the thirteen sets

{ABCK}	{ADGM}	{CEGN}	{AEJP}	
{DEFK}	{BEHM}	{FHAN}	{BFGP}	{KMNP}
{GHJK}	{CFJM}	{BDJN}	{CDHP}	

Set $\{KMNP\} = \ell_\infty$, $A = O$, $B = I$, and $D = I'$. Now coordinatize the nine-point, twelve-line affine plane to obtain

$$
\begin{array}{lll}
A = (0,0) & B = (1,0) & C = (2,0) \\
D = (0,1) & E = (1,1) & F = (2,1) \\
G = (0,2) & H = (1,2) & J = (2,2)
\end{array}
$$

(See Figs. 1.2 and 2.5.) These nine points have the following projective coordinates:

$$
\begin{array}{lll}
A = \langle 0,0,1 \rangle & B = \langle 1,0,1 \rangle & C = \langle 2,0,1 \rangle \\
D = \langle 0,1,1 \rangle & E = \langle 1,1,1 \rangle & F = \langle 2,1,1 \rangle \\
G = \langle 0,2,1 \rangle & H = \langle 1,2,1 \rangle & J = \langle 2,2,1 \rangle
\end{array}
$$

Consider line {CEGN}. The affine equation for {CEG} is $x + y = 2$, or equivalently $x + y + 1 = 0$. The projective equation for {CEGN} is therefore $x + y + z = 0$. Since $N \in \ell_\infty \cap \{CEGN\}$, the projective coordinates of N (in canonical form) are $\langle s, 1, 0 \rangle$ or $\langle 1, 0, 0 \rangle$, and N must satisfy $x + y + z = 0$. Setting $z = 0$ and $y = 1$ gives $s + 1 + 0 = 0$, or $s = 2$. Thus $N = \langle 2, 1, 0 \rangle$.

The point N is also on line {FHAN}. The affine line {FHA} has equation $x + y = 0$, or $x + y + 0 = 0$. Thus {FHAN} has the homogeneous equation $x + y + z \cdot 0 = 0$, or more simply, $x + y = 0$. N must satisfy $x + y = 0$, so setting $y = 1$ gives $x = 2$, and once again $N = \langle 2, 1, 0 \rangle$. It is left as an exercise to see that since {CEG} has affine equation $x + y + 1 = 0$, and {FHA} has affine equation $x + y = 0$, then {BDJ} has affine equation $x + y + 2 = 0$, so {BDJN} has homogeneous equation $x + y + z \cdot 2 = 0$, and $N = \langle 2, 1, 0 \rangle$.

The affine line {ABC} has equation $y = 0$, so {ABCK} has the homogeneous equation $y = 0$ and $K = \langle 1, 0, 0 \rangle$. {CFJM} has equation $x + z = 0$; thus $M = \langle 0, 1, 0 \rangle$. The only possibility left for P is $\langle 1, 1, 0 \rangle$.   ☐

EXERCISES

1. Verify that the point P in the projective plane of order 3 in Example 3 has coordinates $\langle 1, 1, 0 \rangle$. Find the homogeneous equations for the remaining lines in the plane.

The following exercises lead the reader through the coordinatization of the projective plane of order 4 defined in Example 4 below.

☐ **EXAMPLE 4:** **The twenty-one point, twenty-one line projective plane**

$$\mathscr{P}' = \{A, B, C, D, E, F, G, H, I, J, K, L, M, N, O, P, Q, R, S, T, U\}$$

and its lines in $\mathscr{L}'$ consist of

{OIDQF}	{ECPQA}	{QKMJB}	{GHQLN}	{OAJHT}	{TGIPB}
{FTCML}	{EDKNT}	{QRSTU}	{HEIMU}	{UFBAN}	{CGKOU}
{DJLPU}	{RHBCD}	{EFRGJ}	{IARKL}	{MNOPR}	{HFKPS}
{BEOSL}	{JNCIS}	{AGDMS}			

In the exercises below, let O and I be as given, and J = I'. This implies that $\ell'_x = \{OIDFQ\}$ and $\ell'_y = \{OAJHT\}$. Now ℓ_∞ can be selected as any line not containing O, I, or I'. Take ℓ_∞ to be {RHBCD}, and let $(\mathscr{P}, \mathscr{L})$ be the affine plane obtained by removing ℓ_∞ and all its points from $(\mathscr{P}', \mathscr{L}')$. ☐

2. a. Where does ℓ_∞ intersect the axes ℓ_x and ℓ_y?

 b. List the points and lines of $(\mathscr{P}, \mathscr{L})$, grouping the lines in pencils of parallel lines.

Recall that $GF(4)$ is the four-element field $\mathbb{Z}_2[u]/(u^2 + u + 1)$.

3. Find the images of the points of ℓ_x under the isomorphism $\phi : \ell_x \to GF(4)$ described by $\phi(u) = F$.

4. Find Q' and F' on ℓ_y and give affine coordinates to the points of ℓ_x and ℓ_y.

5. Finish the coordinatization of the affine plane $(\mathscr{P}, \mathscr{L})$.

6. Find the homogeneous coordinates of the points in $\mathscr{P}$ (now of the form $(x, y) \in GF(4) \times GF(4)$).

7. Find the homogeneous equations for the lines other than ℓ_∞ in $\mathscr{L}'$.

8. Find the homogeneous coordinates for the points on ℓ_∞.

5.4 PROJECTIVE CONICS

A line in the projective plane $PG_2(F)$ satisfies a homogeneous linear equation

$$p(\mathbf{x}) = a_1 x_1 + a_2 x_2 + a_3 x_3 = 0. \tag{5.2}$$

To reduce the number of variables, here and in the next section, we use a letter such as $\mathbf{x}$ to represent the vector (x_1, x_2, x_3). The expression $p(\mathbf{x})$, or $a_1 x_1 + a_2 x_2 + a_3 x_3$, is called a homogeneous *linear form*. Using determinants, we can find a homogeneous linear equation for the line containing $\langle \mathbf{u} \rangle$ and $\langle \mathbf{v} \rangle$. It is given by

$$\begin{vmatrix} \mathbf{x} \\ \mathbf{u} \\ \mathbf{v} \end{vmatrix} = \begin{vmatrix} x_1 & x_2 & x_3 \\ u_1 & u_2 & u_3 \\ v_1 & v_2 & v_3 \end{vmatrix} = 0.$$

(See Exercises 2 and 3 of Section 5.2.) Thus it is easy to verify when the projective points $\langle \mathbf{u} \rangle$, $\langle \mathbf{v} \rangle$, and $\langle \mathbf{w} \rangle$ are collinear; this occurs exactly when the determinant

$$\begin{vmatrix} \mathbf{u} \\ \mathbf{v} \\ \mathbf{w} \end{vmatrix} = 0.$$

Dually, the lines $\langle \mathbf{u} \rangle$ and $\langle \mathbf{v} \rangle$ intersect in the point $\langle \mathbf{p} \rangle$ where the p_i are given by the determinants

$$p_1 = \begin{vmatrix} u_2 & u_3 \\ v_2 & v_3 \end{vmatrix}, \quad p_2 = \begin{vmatrix} u_3 & u_1 \\ v_3 & v_1 \end{vmatrix}, \quad p_3 = \begin{vmatrix} u_1 & u_2 \\ v_1 & v_2 \end{vmatrix}.$$

There is not a good theory of determinants over division rings; thus the results in this section and the section that follows apply only to projective planes over fields.

When passing to the affine plane over F, the projective point $\langle \mathbf{x} \rangle = \langle x_1, x_2, x_3 \rangle$ ($x_3 \neq 0$) becomes $(x_1/x_3, x_2/x_3)$ which we will denote (X_1, X_2). Recall that the projective points $\langle \mathbf{x} \rangle$ for which $x_3 = 0$ are not in the affine plane but on the line at infinity.

Equation (5.2) becomes $a_1 X_1 + a_2 X_2 + a_3 = 0$. Similarly the equation for a conic in the affine plane over F is given by a quadratic equation

$$f_1 X_1^2 + f_2 X_2^2 + f_3 X_1 + f_4 X_2 + f_5 X_1 X_2 + f_6 = 0. \tag{5.3}$$

Therefore in the projective plane over F, a *conic* is a set of points satisfying a homogeneous quadratic equation $Q(x) = f_1 x_1^2 + f_2 x_2^2 + f_3 x_1 x_3 + f_4 x_2 x_3 + f_5 x_1 x_2 + f_6 x_3^2 = 0$. $Q(\mathbf{x})$ is called a *homogeneous quadratic form*. Note that $Q(\mathbf{x})$ and any of its nonzero multiples define the same conic.

If Q is the product of two (possibly identical) linear forms, then the conic it defines is *degenerate* and consists of a single line or a pair of intersecting lines. If $Q(\mathbf{x})$ is irreducible, the conic is *nondegenerate*. Conics are preserved by nonsingular linear transformations since $(Q \circ T)(\mathbf{x})$ is a homogeneous quadratic form whenever Q is, and $Q \circ T$ is irreducible if and only if Q is irreducible. (See Exercise 3.) In what follows we will assume that all conics are nondegenerate.

In case $F = \mathbb{R}$, the real numbers, more can be said. The projective conic $Q(\mathbf{x}) = 0$ intersects the line at infinity ($x_3 = 0$) in the points which satisfy $f_1 x_1^2 + f_2 x_2^2 + f_5 x_1 x_2 = 0$, which correspond to the affine points satisfying

$$f_1 X_1^2 + f_2 X_2^2 + f_5 X_1 X_2 = 0. \tag{5.4}$$

The left side of (5.4) factors into

$$\frac{1}{2f_1} \left[2f_1 X_1 + \left(f_5 + \sqrt{f_5^2 - 4f_1 f_2} \right) X_2 \right] \left[2f_1 X_1 + \left(f_5 - \sqrt{f_5^2 - 4f_1 f_2} \right) X_2 \right].$$

One of three things happens:

1. If $f_5^2 - 4f_1 f_2 < 0$, then (5.4) has no real solutions except $(0, 0)$, and the projective conic contains no points at infinity. Thus it is completely contained in the affine part of the plane.
2. If $f_5^2 - 4f_1 f_2 = 0$, then (5.4) has as solutions $\{(X_1, X_2) : 2f_1 X_1 + f_5 X_2 = 0\}$, a line through the origin of $\mathbb{R}^2$. Thus the projective conic $Q(\mathbf{x}) = 0$ intersects the line at infinity at a single point, namely the point at infinity on the projective line $2f_1 x_1 + f_5 x_2 = 0$, which is $\langle 1, -f_5/2f_1, 0 \rangle$.
3. If $f_5^2 - 4f_1 f_2 = D > 0$, then the solutions of (5.4) are the points on the pair of lines with equations $2f_1 X_1 + (f_5 \pm \sqrt{D})X_2 = 0$, a pair of lines through $(0, 0)$. In this case the projective conic intersects $x_3 = 0$ in two points, the intersections of the projective lines with equations $2f_1 x_1 + (f_5 \pm \sqrt{D})x_2 = 0$ with the line at infinity, $x_3 = 0$.

The real number $D = f_5^2 - 4f_1 f_2$ is called the *discriminant* of the real affine conic whose equation is given in (5.3). If $D < 0$, the conic is an *ellipse*; if $D = 0$, it is a *parabola*; and if $D > 0$, it is a *hyperbola*. If $D = 0$, the line with equation $2f_1 X_1 + f_5 X_2$ is the *axis* of the parabola; if $D > 0$, the lines with equations $2f_1 X_1 + (f_5 \pm \sqrt{D})X_2 = 0$ are the *asymptotes* of the hyperbola. In Fig. 5.3 the case of the parabola is illustrated.

Any nonsingular linear transformation on $\mathbb{R}^2$ maps parabolas to parabolas, hyperbolas to hyperbolas, and ellipses to ellipses. In the projective plane there is only one type of conic, in the sense that ellipses can be mapped to hyperbolas and parabolas, and, conversely, by linear transformations. To illustrate, suppose that $Q(\mathbf{x}) = 0$ defines an ellipse in the real projective plane. Take any line which intersects that ellipse in two points,

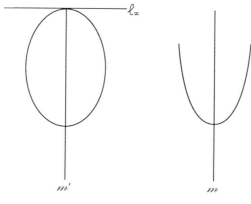

Figure 5.3

$\langle \mathbf{u} \rangle$ and $\langle \mathbf{v} \rangle$. Let T be the linear transformation on $\mathbb{R}^3$ which maps $\mathbf{u}$ to $(1, 0, 0)$, $\mathbf{v}$ to $(0, 1, 0)$, and fixes $(0, 0, 1)$. Then T maps the projective line containing $\langle \mathbf{u} \rangle$ and $\langle \mathbf{v} \rangle$ to the line at infinity; hence T maps the ellipse to a hyperbola. Similarly there is only one type of conic in the projective plane over an arbitrary field.

EXERCISES

1. Determine which of the following equations define ellipses, hyperbolas, and parabolas in the Euclidean plane. If the conic is a parabola, find its axis; if it is a hyperbola, find its asymptotes. In each case find the intersection of the corresponding projective conic with the line at infinity.
   a. $2x^2 + 2y^2 + 4xy - 10x = 0$.
   b. $2x^2 - 3xy + y^2 = 5x + 2y - 6$.
   c. $3xy = 15$.
   d. $(x/a)^2 + (y/b)^2 = 1$ (a, b nonzero real numbers).
   e. $(x/a)^2 - (y/b)^2 = 1$ (a, b nonzero real numbers).
   f. $2x^2 + 2y^2 - 3y + x - 1 = 0$.

2. Use the quadratic formula to verify the factorization of (5.4) given in the text. Recall that a *linear transformation* from F^3 to itself is a mapping $T : F^3 \to F^3$ such that $T(\mathbf{x} + \mathbf{y}) = T(\mathbf{x}) + T(\mathbf{y})$, and $T(a\mathbf{x}) = aT(\mathbf{x})$ for all $\mathbf{x}, \mathbf{y}$, in F^3 and a in F. T is nonsingular if T has an inverse (i.e., if T is one-one and onto).

3. a. Suppose that $Q(\mathbf{x})$ is a homogeneous quadratic form over a field F, and let T be a nonsingular linear transformation from F^3 to itself. Prove that $(Q \circ T)(\mathbf{x})$ is a homogeneous quadratic form over F.
   b. Prove that $(Q \circ T)$ is irreducible if and only if Q is irreducible.

4. Write the equation for each of the projective conics corresponding to the affine conics given in the first three parts of Exercise 1, using the variables x, y, and z. For each parabola find linear transformations of $\mathbb{R}^3$ mapping it into an ellipse and a hyperbola; do the same for each ellipse and each hyperbola.

5.5 PASCAL'S THEOREM

In this section we will show that in a projective plane over a field any five points, no four of which are collinear, lie on one and only one conic. In addition, we will present a generalization of Pappus' Theorem (i.e., Pascal's Theorem which concerns conics in projective space), and show that it holds in a projective plane if and only if the coordinatizing ring is commutative.

Any five collinear points are on all degenerate conics that contain their line, and any five points, four of which are collinear on line ℓ, are on all degenerate conics containing ℓ. However, any five points, no four of which are collinear determine a unique conic in $PG_2(F)$. To show this, the following lemma about determinants is needed.

■ **LEMMA.** If $\langle 1,0,0 \rangle$, $\langle 0,1,0 \rangle$, $\langle 0,0,1 \rangle$, $\langle 1,1,1 \rangle$, and $\langle e_1, e_2, e_3 \rangle$ are distinct points in $PG_2(F)$, then at least two of the following determinants are nonzero:

$$\begin{vmatrix} 1 & 1 \\ e_1 e_2 & e_1 e_3 \end{vmatrix} \quad \begin{vmatrix} 1 & 1 \\ e_1 e_2 & e_2 e_3 \end{vmatrix} \quad \begin{vmatrix} 1 & 1 \\ e_1 e_3 & e_2 e_3 \end{vmatrix}$$

Proof: Without loss of generality, assume that the first two determinants are 0. Then $e_1 e_2 = e_1 e_3$ and $e_1 e_2 = e_2 e_3$.

If $e_1 = 0$, then either e_2 or e_3 is 0. Consequently all three determinants are 0, and $\langle e \rangle = \langle 0,1,0 \rangle$ or $\langle 0,0,1 \rangle$. Assume that $e_1 \neq 0$, and $e_2 = e_3$. Then $e_2 \neq 0$; otherwise, $\langle e_1, e_2, e_3 \rangle = \langle 1,0,0 \rangle$. But $e_2 \neq 0$ implies that $e_1 = e_3 \,(= e_2)$ and that $\langle e_1, e_2, e_3 \rangle = \langle 1,1,1 \rangle$, a contradiction. ■

■ **THEOREM 7.** Let $\langle \mathbf{a} \rangle$, $\langle \mathbf{b} \rangle$, $\langle \mathbf{c} \rangle$, $\langle \mathbf{d} \rangle$, and $\langle \mathbf{e} \rangle$ be five points in $PG_2(F)$, no four of which are collinear. Then there is a unique homogeneous form Q which they satisfy.

Proof: It may be assumed that no three of $\langle \mathbf{a} \rangle$, $\langle \mathbf{b} \rangle$, $\langle \mathbf{c} \rangle$, and $\langle \mathbf{d} \rangle$ are collinear. By making a linear transformation, we may take these points to be $\langle 1,0,0 \rangle$, $\langle 0,1,0 \rangle$, $\langle 0,0,1 \rangle$, and $\langle 1,1,1 \rangle$, with $\langle \mathbf{e} \rangle$ distinct from all of them. We must find $f_1, \ldots, f_6 \in F$ so that $Q(\mathbf{x}) = f_1 x_1^2 + f_2 x_2^2 + f_3 x_1 x_3$ $+ f_4 x_2 x_3 + f_5 x_1 x_2 + f_6 x_3^2 = 0$ is satisfied by $\langle \mathbf{x} \rangle = \langle 1,0,0 \rangle$, $\langle 0,1,0 \rangle$, $\langle 0,0,1 \rangle$, $\langle 1,1,1 \rangle$, and $\langle \mathbf{e} \rangle$. Substituting $\langle 1,0,0 \rangle$, $\langle 0,1,0 \rangle$, and $\langle 0,0,1 \rangle$ into $Q(\mathbf{x})$ gives

$$f_1 = f_2 = f_6 = 0.$$

Substituting $\langle 1, 1, 1 \rangle$ gives

$$f_3 + f_4 + f_5 = 0.$$

Finally, substituting $\langle \mathbf{e} \rangle$, we have

$$f_3 e_1 e_2 + f_4 e_1 e_3 + f_5 e_2 e_3 = 0.$$

By the lemma, we can assume, without loss of generality, that

$$\begin{vmatrix} 1 & 1 \\ e_1 e_2 & e_1 e_3 \end{vmatrix} \neq 0.$$

Thus

$$\begin{pmatrix} 1 & 1 \\ e_1 e_2 & e_1 e_3 \end{pmatrix} \begin{pmatrix} y_3 \\ y_4 \end{pmatrix} = \begin{pmatrix} -1 \\ -e_2 e_3 \end{pmatrix}$$

has a unique solution $\begin{pmatrix} f_3 \\ f_4 \end{pmatrix}$ in $F \times F$. Hence $f_3 x_1 x_2 + (-1 - f_3) x_1 x_3 + x_2 x_3 = 0$ is the unique homogeneous quadratic equation satisfied by $\langle \mathbf{x} \rangle = \langle 1, 0, 0 \rangle$, $\langle 0, 1, 0 \rangle$, $\langle 0, 0, 1 \rangle$, $\langle 1, 1, 1 \rangle$, and $\langle \mathbf{e} \rangle$. ∎

■ **THEOREM 8. Pascal's Theorem.** Let $\langle \mathbf{a} \rangle$, $\langle \mathbf{b} \rangle$, $\langle \mathbf{c} \rangle$, $\langle \mathbf{d} \rangle$, $\langle \mathbf{e} \rangle$, and $\langle \mathbf{f} \rangle$ be six points, no four of which are collinear, in $PG_2(F)$. Then $\ell(\langle \mathbf{a} \rangle, \langle \mathbf{b} \rangle) \cap \ell(\langle \mathbf{d} \rangle, \langle \mathbf{e} \rangle)$, $\ell(\langle \mathbf{b} \rangle, \langle \mathbf{c} \rangle) \cap \ell(\langle \mathbf{e} \rangle, \langle \mathbf{f} \rangle)$, and $\ell(\langle \mathbf{c} \rangle, \langle \mathbf{d} \rangle) \cap \ell(\langle \mathbf{f} \rangle, \langle \mathbf{a} \rangle)$ are collinear if and only if there is a homogeneous quadratic form $Q(\mathbf{x})$ satisfied by $\langle \mathbf{a} \rangle, \ldots, \langle \mathbf{f} \rangle$. (See Fig. 5.4.)

Proof:

Case 1. Suppose that no three of $\langle \mathbf{a} \rangle$, $\langle \mathbf{b} \rangle$, $\langle \mathbf{c} \rangle$, $\langle \mathbf{d} \rangle$, are collinear, and as before, assume that they are $\langle 1, 0, 0 \rangle$, $\langle 0, 1, 0 \rangle$, $\langle 0, 0, 1 \rangle$, and $\langle 1, 1, 1 \rangle$,

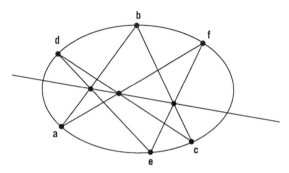

Figure 5.4

respectively. Then

$$\ell(\langle \mathbf{a}\rangle, \langle \mathbf{b}\rangle) \cap \ell(\langle \mathbf{d}\rangle, \langle \mathbf{e}\rangle) = \langle(e_3 - e_1, e_3 - e_2, 0)\rangle,$$

$$\ell(\langle \mathbf{b}\rangle, \langle \mathbf{c}\rangle) \cap \ell(\langle \mathbf{e}\rangle, \langle \mathbf{f}\rangle) = \langle(0, e_2 f_1 - e_1 f_2, e_3 f_1 - e_1 f_3)\rangle,$$

$$\ell(\langle \mathbf{c}\rangle, \langle \mathbf{d}\rangle) \cap \ell(\langle \mathbf{f}\rangle, \langle \mathbf{a}\rangle) = \langle f_2, f_2, f_3\rangle.$$

These points are collinear if and only if the following determinant is zero:

$$\begin{vmatrix} e_3 - e_1 & e_3 - e_2 & 0 \\ 0 & e_2 f_1 - e_1 f_2 & e_3 f_1 - e_1 f_3 \\ f_2 & f_2 & f_3 \end{vmatrix}$$

Expanding this determinant gives

$$(e_3 - e_1)e_2 f_1 f_3 + (e_1 - e_2)e_3 f_1 f_2 + (e_2 - e_3)e_1 f_2 f_3,$$

which equals 0 if and only if f satisfies

$$(e_3 - e_1)e_2 x_1 x_3 + (e_1 - e_2)e_3 x_1 x_2 + (e_2 - e_3)e_1 x_2 x_3 = 0,$$

a homogeneous quadratic equation satisfied by $\langle \mathbf{a}\rangle, \ldots, \langle \mathbf{e}\rangle$.

CASE 2. If some three of $\langle \mathbf{a}\rangle, \langle \mathbf{b}\rangle, \langle \mathbf{c}\rangle, \langle \mathbf{d}\rangle$, are collinear, satisfying a homogeneous linear form $p(x) = 0$, and if all six points satisfy a homogeneous quadratic equation $Q(\mathbf{x}) = 0$, then, by Theorem 7, Q is unique and of the form $p(\mathbf{x})r(\mathbf{x}) = 0$, where $r(\mathbf{x})$ is a linear form. This means that the six points are on two lines. The proof breaks further into cases, depending on which three of $\langle \mathbf{a}\rangle, \langle \mathbf{b}\rangle, \langle \mathbf{c}\rangle, \langle \mathbf{d}\rangle$, are collinear, of which the following case is typical. If $\langle \mathbf{a}\rangle, \langle \mathbf{b}\rangle, \langle \mathbf{d}\rangle$, are collinear, then $\langle \mathbf{c}\rangle, \langle \mathbf{e}\rangle$, and $\langle \mathbf{f}\rangle$ are collinear. This implies that

$$\ell(\langle \mathbf{a}\rangle, \langle \mathbf{b}\rangle) \cap \ell(\langle \mathbf{d}\rangle, \langle \mathbf{e}\rangle) = \langle \mathbf{d}\rangle,$$

$$\ell(\langle \mathbf{b}\rangle, \langle \mathbf{c}\rangle) \cap \ell(\langle \mathbf{e}\rangle, \langle \mathbf{f}\rangle) = \langle \mathbf{c}\rangle,$$

$$\ell(\langle \mathbf{c}\rangle, \langle \mathbf{d}\rangle) \cap \ell(\langle \mathbf{f}\rangle, \langle \mathbf{a}\rangle) \in \ell(\langle \mathbf{c}\rangle, \langle \mathbf{d}\rangle).$$

Conversely, if $\ell(\langle \mathbf{a}\rangle, \langle \mathbf{b}\rangle) \cap \ell(\langle \mathbf{d}\rangle, \langle \mathbf{e}\rangle)$, $\ell(\langle \mathbf{b}\rangle, \langle \mathbf{c}\rangle) \cap \ell(\langle \mathbf{e}\rangle, \langle \mathbf{f}\rangle)$, and $\ell(\langle \mathbf{c}\rangle, \langle \mathbf{d}\rangle) \cap \ell(\langle \mathbf{f}\rangle, \langle \mathbf{a}\rangle)$ are collinear, again the proof breaks down into cases depending on which three of $\langle \mathbf{a}\rangle, \langle \mathbf{b}\rangle, \langle \mathbf{c}\rangle, \langle \mathbf{d}\rangle$, are collinear; once more a typical case is presented. If $\langle \mathbf{a}\rangle, \langle \mathbf{b}\rangle, \langle \mathbf{d}\rangle$, are collinear, satisfying $p(\mathbf{x}) = 0$, then $\ell(\langle \mathbf{a}\rangle, \langle \mathbf{b}\rangle) \cap \ell(\langle \mathbf{d}\rangle, \langle \mathbf{e}\rangle) = \langle \mathbf{d}\rangle$, and the line containing $\langle \mathbf{d}\rangle$ and $\ell(\langle \mathbf{c}\rangle, \langle \mathbf{d}\rangle) \cap \ell(\langle \mathbf{f}\rangle, \langle \mathbf{a}\rangle)$ is $\ell(\langle \mathbf{c}\rangle, \langle \mathbf{d}\rangle)$. Thus $\ell(\langle \mathbf{b}\rangle, \langle \mathbf{c}\rangle) \cap \ell(\langle \mathbf{e}\rangle, \langle \mathbf{f}\rangle) \in \ell(\langle \mathbf{c}\rangle, \langle \mathbf{d}\rangle)$. But this implies that $\ell(\langle \mathbf{b}\rangle, \langle \mathbf{c}\rangle) \cap$

$\ell(\langle\mathbf{e}\rangle,\langle\mathbf{f}\rangle) = \langle\mathbf{c}\rangle$, so $\langle\mathbf{c}\rangle$, $\langle\mathbf{e}\rangle$, and $\langle\mathbf{f}\rangle$ are collinear and satisfy $r(\mathbf{x}) = 0$ for some homogeneous linear form r. Thus all six points satisfy $p(\mathbf{x})r(\mathbf{x}) = 0$.

∎

Earlier in Section 3.7 and Section 4.6, we discussed Pappus' Theorem for the affine plane; its configuration was shown in Fig. 4.12. Upon adjoining a line at infinity, the theorem can be stated projectively as follows.

■ **PAPPUS' THEOREM FOR THE PROJECTIVE PLANE.** Let $\langle\mathbf{a}\rangle$, $\langle\mathbf{b}\rangle$, $\langle\mathbf{c}\rangle$, $\langle\mathbf{d}\rangle$, $\langle\mathbf{e}\rangle$, and $\langle\mathbf{f}\rangle$ be distinct points with $\langle\mathbf{a}\rangle \in \ell(\langle\mathbf{e}\rangle,\langle\mathbf{c}\rangle)$ and $\langle\mathbf{b}\rangle \in \ell(\langle\mathbf{d}\rangle,\langle\mathbf{f}\rangle)$. Then, if $\ell(\langle\mathbf{a}\rangle,\langle\mathbf{b}\rangle) \cap \ell(\langle\mathbf{d}\rangle,\langle\mathbf{e}\rangle) \in \ell_\infty$ and $\ell(\langle\mathbf{f}\rangle,\langle\mathbf{e}\rangle) \cap \ell(\langle\mathbf{b}\rangle,\langle\mathbf{c}\rangle) \in \ell_\infty$, it follows that $\ell(\langle\mathbf{c}\rangle,\langle\mathbf{d}\rangle) \cap \ell(\langle\mathbf{f}\rangle,\langle\mathbf{a}\rangle) \in \ell_\infty$.

Since any line could be chosen for ℓ_∞, Pappus' Theorem can be stated more generally as follows:

■ **PAPPUS' THEOREM FOR THE PROJECTIVE PLANE.** Let $\langle\mathbf{a}\rangle$, $\langle\mathbf{b}\rangle$, $\langle\mathbf{c}\rangle$, $\langle\mathbf{d}\rangle$, $\langle\mathbf{e}\rangle$, and $\langle\mathbf{f}\rangle$ be distinct points with $\langle\mathbf{a}\rangle \in \ell(\langle\mathbf{e}\rangle,\langle\mathbf{c}\rangle)$ and $\langle\mathbf{b}\rangle \in \ell(\langle\mathbf{d}\rangle,\langle\mathbf{f}\rangle)$. Then $\ell(\langle\mathbf{a}\rangle,\langle\mathbf{b}\rangle) \cap \ell(\langle\mathbf{d}\rangle,\langle\mathbf{e}\rangle)$, $\ell(\langle\mathbf{f}\rangle,\langle\mathbf{e}\rangle) \cap \ell(\langle\mathbf{b}\rangle,\langle\mathbf{c}\rangle)$, and $\ell(\langle\mathbf{c}\rangle,\langle\mathbf{d}\rangle) \cap \ell(\langle\mathbf{f}\rangle,\langle\mathbf{a}\rangle)$ are collinear points.

The conic here is degenerate, consisting of the lines $\ell(\langle\mathbf{e}\rangle,\langle\mathbf{c}\rangle)$ and $\ell(\langle\mathbf{d}\rangle,\langle\mathbf{f}\rangle)$, and Pappus' Theorem is a degenerate case of Pascal's Theorem, so that any projective plane in which Paschal's Theorem holds satisfies Pappus' Theorem as well. Furthermore, if a projective plane satisfies Paschal's Theorem, the affine plane obtained by removing any one of its lines satisfies Pappus' Theorem and can therefore be coordinatized by a field. (See Theorem 18 of Chapter 4.) Thus we have the following equivalences for an arbitrary division ring.

■ **THEOREM 9.** For $\mathbb{D}$ any division ring, the following statements are equivalent.

    i. $\mathbb{D}$ is a field.
    ii. Paschal's Theorem holds in the projective plane over $\mathbb{D}$.
    iii. Pappus' Theorem holds in the affine plane over $\mathbb{D}$.

EXERCISES

1. Prove the remaining cases of Theorem 8.
2. For each set of five points given below, find an equation for the unique conic containing those points:
   a. $\langle 1,0,1\rangle$, $\langle 0,0,1\rangle$, $\langle 2,1,4\rangle$, $\langle 6,1,7\rangle$, $\langle\pi,5,6\rangle$, in the real projective plane.
   b. $\langle 1,0,1\rangle$, $\langle 0,0,1\rangle$, $\langle 2,1,4\rangle$, $\langle 6,1,7\rangle$, $\langle 1,5,6\rangle$, in the rational projective plane.
   c. $\langle 3,3,4\rangle$, $\langle 2,1,4\rangle$, $\langle 5,6,1\rangle$, $\langle 1,0,1\rangle$, $\langle 1,2,2\rangle$, in the projective plane over $\mathbb{Z}_7$.

5.6 NON-DESARGUESIAN COORDINATE PLANES

There is a vast amount of literature on non-Desarguesian projective planes, with results both more numerous and deeper than can be presented here. The books listed in the references at the end of the chapter all deal at great length with non-Desarguesian projective planes, and there are still many open questions in the area—for example, is there a finite projective plane whose order is not a prime power? Here some finite non-Desarguesian projective planes will be described using coordinate systems that are not fields and, in fact, not even rings. In the next section some specific examples will be given, and in section 5.8 two nonisomorphic coordinate systems that define isomorphic projective planes will be described.

Not all of the properties of division rings are necessary to construct a coordinate projective plane as described in Theorem 3 and its corollary in Section 5.2. Here we pick out a minimal set of algebraic properties, describe the construction of the associated projective coordinate planes, and give some examples of specific systems. If the algebraic system is not a division ring, its coordinate plane need not be Desarguesian. The material in this section and in the next section was adapted from Veblen-Wedderburn (1907).

Definition. A *Veblen-Wedderburn system* is a triple $(\mathscr{S}, +, \cdot)$ satisfying the following conditions:

*VW*1. $(\mathscr{S}, +)$ is an abelian group with identity 0.

*VW*2. $\mathscr{S}$ is closed under multiplication.

*VW*3. Given a and b with $a \neq 0$, there are unique elements d and d' such that $da = b$ and $ad' = b$.

*VW*4. $0 \cdot a = a \cdot 0 = 0$ for all a in $\mathscr{S}$.

*VW*5. $c(a + b) = ca + cb$ for all $a, b, c \in \mathscr{S}$.

Given a Veblen-Wedderburn system fix two (not necessarily distinct) nonzero elements ϕ and ψ. Define $\mathscr{P}'$ and $\mathscr{L}'$ as follows:

$$\mathscr{P}' = \{(p, q, \phi) : p, q \in \mathscr{S}\} \cup \{(p, \phi, 0) : p \in \mathscr{S}\} \cup \{(\phi, 0, 0)\}.$$

Points in the first set will be referred to as "type 1 points," those in the second will be referred to as "type 2 points," while $(\phi, 0, 0)$ is the only type 3 point.

A *line* is a set of points that satisfies one of the following types of equations where b and c take on arbitrary values in $\mathcal{S}$:

$$x\psi + yb + zc = 0 \qquad \text{(Type 1 line)},$$

$$y\psi + zc = 0 \qquad \text{(Type 2 line)},$$

$$z\psi = 0 \qquad \text{(Type 3 line)}.$$

The set of lines is denoted by $\mathcal{L}'$. Following the notation introduced with the corollary to Theorem 3, the lines have the forms $[\psi, b, c]$, $[0, \psi, c]$, and $[0, 0, \psi]$. Since the line $[0, 0, 1]$ in the Desarguesian projective plane over a division ring is the usual line at infinity, the type 3 line $[0, 0, \psi]$ is referred to as the line at infinity as well.

In contrast to projective planes over division rings in which both distributive laws hold, there is no reason to assume that for a nonzero element $d \in \mathcal{S}$, the points that satisfy $x\psi + yb + zc = 0$ will satisfy its multiple $x\psi d + ybd + zcd = 0$. Thus general equations of the form $xa + yb + zc = 0$ are not equations for lines unless they are of type 1, 2, or 3 above. If $\mathcal{S}$ were a division ring, and $\phi = \psi = 1$, the definitions of $\mathcal{P}'$ and $\mathcal{L}'$ above would give the usual projective coordinate plane over $\mathcal{S}$. It will next be shown that if $\mathcal{S}$ is finite, then $(\mathcal{P}', \mathcal{L}')$ is a projective plane. If $\mathcal{S}$ is infinite, right distributivity is needed to show that each pair of points is on some line.

In the case where $\mathcal{S}$ is finite and has n elements, both $\mathcal{P}'$ and $\mathcal{L}'$ have $n^2 + n + 1$ elements. But more can be said.

■ **THEOREM 10.** There are $n + 1$ points on each line where $n = |\mathcal{S}|$.

Proof: The equation $z\psi = 0$ is satisfied by the $n + 1$ points of types 2 and 3 only.

Consider the equation $y\psi + zc = 0$. If $c = 0$, the equation becomes $y\psi = 0$, and by $VW3$ and $VW4$, $y = 0$. Thus the equation is satisfied by the n type 1 points of the form $(p, 0, \phi)$, and the type 3 point $(\phi, 0, 0)$.

Suppose that $c \neq 0$. By definition $z = \phi$ or 0. If $z = \phi$, then there is a unique solution y_0 for the equation $y\psi = -\phi c$, and the n points (x, y_0, ϕ) satisfy $y\psi + zc = 0$. Again, using $VW3$ and $VW4$, if $z = 0$, y must be 0, and $(\phi, 0, 0)$ is the $(n + 1)$-st point satisfying $y\phi + zc = 0$.

Showing that lines of type 1 whose equations are of the form $x\psi + yb + zc = 0$ have $n + 1$ points is done similarly and is left as an exercise.

■

■ **THEOREM 11.** There are exactly $n + 1$ lines through each point.

Proof: The point $(\phi, 0, 0)$ is on the $n + 1$ lines whose equations are $y\psi + zc = 0$ and $z\psi = 0$. For the point $(p, \phi, 0)$, solve for $y = b$ the equation $p\psi = -\phi y$. Then $p\psi + \phi b + 0c = 0$ for every value of c. Thus

$(p, \phi, 0)$ is on the n lines whose equations are $x\psi + yb + zc = 0$, as well as the line $z\psi = 0$.

For the point (p, q, ϕ), first solve the equation $q\psi = -\phi z$ for $z = c$. Then $p0 + q\psi + \phi c = 0$, so (p, q, ϕ) is on the line whose equation is $y\psi + zc = 0$. For each value $d \in \mathscr{S}$, solve for $y = b_d$ the equation $qy = -p\psi - \phi d$. Thus $p\psi + qb_d + \phi d = 0$, and (p, q, ϕ) is on the line whose equation is $x\psi + yb_d + zd = 0$, and (p, q, ϕ) is on $n + 1$ lines. ∎

■ **THEOREM 12.** Any two lines have exactly one point in common.

Proof:

CASE 1 (*both lines of type* 1). Suppose that $x\psi + yb + zc = 0$ and $x\psi + yb' + zc' = 0$ are the equations for two type one lines. Since $\mathscr{S}$ is left-distributive, subtracting gives $y(b - b') + z(c - c') = 0$.

If $c = c'$, then $b \neq b'$ (otherwise, the lines would be the same), and we have $y(b - b') = 0$. Thus $y = 0$, and the unique solution $x = p$ of $x\psi = -\phi c$ gives the point of intersection $(p, 0, \phi)$.

If $b = b'$ and $c \neq c'$, then $z(c - c') = 0$, so $z = 0$. If $y = 0$, then $x\psi = 0$, which implies that $x = 0$, a contradiction. Thus $y = \phi$. Solving $x\psi = -\phi b$ for $x = p$ gives the point of intersection $(p, \phi, 0)$.

If $b \neq b'$ and $c \neq c'$, then if $z = 0$, $y(b - b') = 0$. Therefore $y = 0$, and it follows that $x = 0$, a contradiction. Thus $z = \phi$. First find the unique solution $y = y_0$ for $y(b - b') = -\phi(c - c')$. Then find the solution $x = x_0$ for $x\psi = -(y_0 b + \phi c)$. The point (x_0, y_0, ϕ) is the point of intersection.

CASE 2 (*a type* 1 *and a type* 2 *line*). Suppose that the lines have the equations $x\psi + yb + zc = 0$ and $y\psi + zc' = 0$. Then, if $z = 0$, the second equation gives $y = 0$, and it follows from the first equation that $x = 0$, a contradiction. Thus $z = \phi$. Then $y = y_0$, the solution to $y\psi = -\phi c'$, and $x = x_0$, the solution to $x\psi = -(y_0 b + \phi c)$.

CASE 3 (*a type* 1 *and a type* 3 *line*). Here the equations are $x\psi + yb + zc = 0$ and $z\psi = 0$. Thus $z = 0$. If $y = 0$, then $x = 0$, so y cannot be 0. Therefore $y = \phi$, and x is the solution for $x\psi = -\phi b$.

CASE 4 (*two type* 2 *lines*). The equations are $y\psi + zc = 0$ and $y\psi + zc' = 0$ with $c \neq c'$. Thus $z(c - c') = 0$, which implies that $z = 0$. Thus $y = 0$, and $(\phi, 0, 0)$ is the point of intersection.

CASE 5 (*a type* 2 *and a type* 3 *line*). The only solution to $y\psi + zc = 0$ and $z\psi = 0$ is $(\phi, 0, 0)$. ∎

Left-distributivity was used several times in the proof above. To give an analogous proof that any two points are on one and only one line requires *right* distributivity, which does not necessarily hold in $\mathscr{S}$. However, *if* $\mathscr{S}$ *is finite*, the following general result can be used to show that $(\mathscr{P}', \mathscr{L}')$ is a projective plane.

■ **THEOREM 13.** Given a system of $n^2 + n + 1$ points and $n^2 + n + 1$ lines such that every line has $n + 1$ points, every point is on $n + 1$ lines, and every pair of lines meets in a point, then each pair of points is on one and only one line.

Proof: A given pair of points is on at most one line, since two lines meet at only one point. Suppose that P and Q are distinct points. Let $m_1, \ldots, m_{n+1}$ be the $n + 1$ lines through Q, and let ℓ be a line containing P. If Q is in ℓ, we're done. Otherwise, let $R_i = \ell \cap m_i$ $(i = 1, \ldots, n + 1)$. If $i \neq j$, then $R_i \neq R_j$, otherwise Q and R_i $(= R_j)$ would be in $m_i \cap m_j$, which contradicts the statement that each pair of lines meets in only one point. But line ℓ now contains the distinct points $R_1, \ldots, R_{n+1}$ and P. Since ℓ only has $n + 1$ points, $P = R_i$ for some i. Thus $P \in m_i$, and m_i is the line containing P and Q.   ■

The results obtained thus far can be summarized in the following theorem.

■ **THEOREM 14.** If $(\mathscr{S}, +, \cdot)$ is a finite Veblen-Wedderburn system with n elements, and if ϕ and ψ are nonzero elements of $\mathscr{S}$, then $(\mathscr{P}', \mathscr{L}')$ is a projective plane of order n where

$$\mathscr{P}' = \{(x, y, \phi) : x, y \in \mathscr{S}\} \cup \{(x, \phi, 0) : x \in \mathscr{S}\} \cup \{(\phi, 0, 0)\}$$

and $\ell \in \mathscr{L}'$ if and only if ℓ is the set of points satisfying an equation of the form

$$x\psi + yb + zc = 0,$$

$$y\psi + zc = 0,$$

or

$$z\psi = 0.$$

EXERCISES

1. Suppose that $(\mathscr{S}, +, \cdot)$ is a Veblen-Wedderburn system with n elements, and let ϕ and ψ be nonzero elements in $\mathscr{S}$. Let $\mathscr{P}' = \{(p, q, \phi) : p, q \in \mathscr{S}\} \cup \{(p, \phi, 0) : p \in \mathscr{S}\} \cup \{(\phi, 0, 0)\}$. Show that exactly $n + 1$ elements of $\mathscr{P}'$ satisfy each equation of the form $x\psi + yb + zc = 0$.

*2. Prove that the commutativity of addition can be proved from the other postulates for a Veblen-Wedderburn system.

3. Prove that if $\mathscr{S}$ is an infinite Veblen-Wedderburn system in which the distributive law $(a + b)c = ac + bc$ also holds, then in $(\mathscr{P}', \mathscr{L}')$ as defined above, any two points are on one and only one line. (Note that if multiplication is commutative in an infinite Veblen-Wedderburn system, both distributive laws hold, and therefore $(\mathscr{P}', \mathscr{L}')$ is a projective plane.)

*__4.__ Prove that if $(\mathcal{T}, +, \cdot)$ is a finite system that satisfies the first four axioms for a Veblen-Wedderburn system, together with right-distributivity

$$VW5' \cdot (a + b)c = ac + bc,$$

then $(\mathcal{P}', \mathcal{L}')$ is a projective plane, where

$$\mathcal{P}' = \{(p, q, \phi) : p, q \in \mathcal{T}\} \cup \{(p, \phi, 0) : p \in \mathcal{T}\} \cup \{(\phi, 0, 0)\},$$

$$\mathcal{L}' = \{[\psi, b, c] : b, c \in \mathcal{T}\} \cup \{[0, \psi, c] : c \in \mathcal{T}\} \cup \{[0, 0, \psi]\},$$

for nonzero elements ϕ and ψ of $\mathcal{T}$.

__5.__ Show that if $(\mathcal{S}, +, \cdot)$ is a Veblen-Wedderburn system, then $(\mathcal{S}, +, \circ)$ satisfies $VW1$–$VW4$, together with right-distributivity, where $a \circ b$ is defined by

$$a \circ b = b \cdot a.$$

5.7 SOME EXAMPLES OF VEBLEN-WEDDERBURN SYSTEMS

Some Veblen-Wedderburn systems are based on finite and infinite fields. We first give a general construction and then show some specific applications, which include an infinite non-Desarguesian plane.

Suppose that F is a field of characteristic different from 2. Define $\mathcal{S}$ to be $\{a + di + ej : a, d, e \in F\}$, where i and j are distinct symbols. Let $x^3 = bx + c$ be a cubic polynomial, irreducible over F. Define addition by $(a + di + ej) + (a' + d'i + e'j) = (a + a') + (d + d')i + (e + e')j$, the addition being done in F. Elements are multiplied as polynomials, with the products of i's and j's given as follows:

$\times$	i	j
i	j	$c + bi$
j	$c + bi$	$-b^2 - 8ci - 2bj$

Then $(\mathcal{S}, +, \times)$ is a Veblen-Wedderburn system.

☐__EXAMPLE 5:__ Take F to be $\mathbb{Z}_3$, and take the irreducible cubic to be $x^3 = 1 + x$. Then $\mathcal{S} = \{a + di + ej : a, d, e \in \mathbb{Z}_3\}$, and the multiplication table for i and j is as follows.

$\times$	i	j
i	j	$1 + i$
j	$1 + i$	$-1 + i + j$

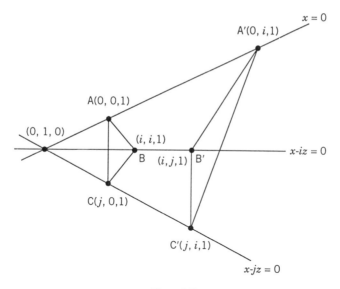

Figure 5.5

The projective plane over $\mathscr{S}$ is of order 27, thus there are 757 points, 757 lines, and 28 points on each line. □

☐**EXAMPLE 6:** Let $F = \mathbb{Q}$, the rational numbers. The equation $x^3 = 2$ has no solution, so $c = 2$ and $b = 0$. The table for multiplication of i and j is the following.

×	i	j
i	j	2
j	2	$-16i$

Desargues' Theorem does not hold in the infinite coordinate plane given by this example, as Fig. 5.5 shows.

It is not difficult to verify that the following are the equations for the indicated lines:

$$\ell(A, C) : y = 0; \quad \ell(A, B) : x - y = 0; \quad \ell(A', C') : y - iz = 0;$$

$$\ell(A', B') : x + \frac{(1 + i + j)y}{17} - \frac{(2 + i + j)z}{17} = 0;$$

$$\ell(B, C) : x - (1 - i)y - jz = 0;$$

$$\ell(B', C') : x + y - (i + j)z = 0;$$

$D = \ell(A, C) \cap \ell(A', C') = (1, 0, 0)$,

$E = \ell(A, B) \cap \ell(A', B') = (s, s, 1)$ where $s = \dfrac{8[(17/8) + 2i + j]}{177}$,

$F = \ell(B, C) \cap \ell(B', C') = \left(\dfrac{-1 + i + 2j}{3}, \dfrac{1 + 2i + j}{3}, 1 \right)$.

The line $\ell(D, E)$ has the equation

$$y - \frac{8[(17/8) + 2i + j]}{177} z = 0,$$

and F does not satisfy this equation.   □

□**EXAMPLE 7:** Our final example has the same elements and addition table as $\mathbb{Z}_5$, with multiplication given as

×	0	1	2	3	4
0	0	0	0	0	0
1	0	1	2	3	4
2	0	4	3	2	1
3	0	3	1	4	2
4	0	2	4	1	3

Here take $\phi = \psi = 1$. Although $\mathscr{S}$ is not isomorphic to $\mathbb{Z}_5$, there is a collineation between the plane over $\mathbb{Z}_5$ and the plane over $\mathscr{S}$. (Recall that such a collineation is a one-one onto mapping from the points of the plane over $\mathbb{Z}_5$ to the plane over $\mathscr{S}$ which takes collinear points to collinear points.)   □

EXERCISES

1. Show that the system $(\mathscr{S}, +, \times)$ where F is a field, $x^3 = bx + c$ is an irreducible polynomial over F and $\mathscr{S}$ is defined by

$$\mathscr{S} = \{a + di + ej : a, d, e \in F\}$$

is a Veblen-Wedderburn system, where addition and multiplication are defined as for polynomials in i and j, with $ij = ji = c + bi$ and $i^2 = j$, and $j^2 = -b^2 - 8ci - 2bj$.

2. Give an example of a Veblen-Wedderburn system with 125 elements. Give an example with $7^3 = 343$ elements.

3. Give an example of two triangles that are perspective from a point in the plane of Example 5. Are they perspective from a line?

4. Show that $[(i + 1)i]j \neq (i + 1)(ij)$ in the system given in Example 5.

*5. Show that the plane of Example 7 is isomorphic to the plane over $\mathbb{Z}_5$. That is, find a bijection between the points of the planes that maps collinear points to collinear points.

5.8 A PROJECTIVE PLANE OF ORDER 9

There are four known nonisomorphic projective planes of order 9, the Desarguesian one coordinatized by $GF(9)$, and three non-Desarguesian ones—the one to be described below, its dual, and the self-dual "Hughes plane." That these are the only four was announced in Lam, Kolesova, and Thiel (1991). Their results also imply that there are exactly seven nonisomorphic *affine* planes of order 9. The projective planes of order 9 are all discussed in detail in Hughes and Piper (1973) and Stevenson (1972). These are the smallest examples of non-Desarguesian projective planes: The only planes of orders 2, 3, 4, 5, 7, and 8 are the Desarguesian ones coordinatized by the appropriate Galois fields, and there is no projective plane of order 6. All Desarguesian planes are self-dual; hence the plane described below is the smallest non-self-dual projective plane.

In this section two nonisomorphic 9-element Veblen-Wedderburn systems will be presented, and it will be shown that they determine isomorphic projective planes. The systems are modifications of two of the four right-distributive systems presented in Hall (1943). Hall in fact proved that all four of his systems coordinatize the same projective plane.

The systems $\mathcal{R}$ and $\mathcal{S}$ each have the same set of elements, namely $\mathcal{R} = \mathcal{S} = \{a + bu : a, b \in \mathbb{Z}_3\}$. In each case addition is the same as the addition in $GF(9)$, given by $(a + bu) + (c + du) = (a + c) + (b + d)u$, the addition being done mod 3. Multiplication is given by the tables for $\circ_{\mathcal{R}}$ and $\circ_{\mathcal{S}}$

$\circ_{\mathcal{R}}$	0	1	2	u	$u + 1$	$u + 2$	$2u$	$2u + 1$	$2u + 2$
0	0	0	0	0	0	0	0	0	0
1	0	1	2	u	$u + 1$	$u + 2$	$2u$	$2u + 1$	$2u + 2$
2	0	2	1	$2u$	$2u + 2$	$2u + 1$	u	$u + 2$	$u + 1$
u	0	u	$2u$	2	$u + 2$	$2u + 2$	1	$u + 1$	$2u + 1$
$u + 1$	0	$u + 1$	$2u + 2$	$2u + 1$	2	u	$u + 2$	$2u$	1
$u + 2$	0	$u + 2$	$2u + 1$	$u + 1$	$2u$	2	$2u + 2$	1	u
$2u$	0	$2u$	u	1	$2u + 1$	$u + 1$	2	$2u + 2$	$u + 2$
$2u + 1$	0	$2u + 1$	$u + 2$	$2u + 2$	u	1	$u + 1$	2	$2u$
$2u + 2$	0	$2u + 2$	$u + 1$	$u + 2$	1	$2u$	$2u + 1$	u	2

$\circ_{\mathcal{S}}$	0	1	2	u	$u + 1$	$u + 2$	$2u$	$2u + 1$	$2u + 2$
0	0	0	0	0	0	0	0	0	0
1	0	1	2	u	$u + 1$	$u + 2$	$2u$	$2u + 1$	$2u + 2$
2	0	2	1	$2u$	$2u + 2$	$2u + 1$	u	$u + 2$	$u + 1$
u	0	u	$2u$	$u + 1$	$2u + 1$	1	$2u + 2$	2	$u + 2$
$u + 1$	0	$u + 1$	$2u + 2$	1	$u + 2$	$2u$	2	u	$2u + 1$
$u + 2$	0	$u + 2$	$2u + 1$	$2u + 2$	1	u	$u + 1$	$2u$	2
$2u$	0	$2u$	u	$u + 2$	2	$2u + 2$	$2u + 1$	$u + 1$	1
$2u + 1$	0	$2u + 1$	$u + 2$	2	$2u$	$u + 1$	1	$2u + 2$	u
$2u + 2$	0	$2u + 2$	$u + 1$	$2u + 1$	u	2	$u + 2$	1	$2u$

■   **THEOREM 15.** Both $(\mathcal{R},+,\circ_{\mathcal{R}})$ and $(\mathcal{S},+,\circ_{\mathcal{S}})$ are Veblen-Wedderburn systems.

Proof:

VW 1. Each of the systems is an abelian group under addition since addition is the same as in the field $GF(9)$.

VW 2. Each is closed under multiplication since the tables contain no elements from outside the system.

VW 3. Given a and b with $a \neq 0$, it must be shown that there are unique elements d and d' with $da = b$ and $ad' = b$. The existence and uniqueness of d follows since each nonzero column contains each element; analogously d' exists and is unique since each nonzero row contains each element.

VW 4. In each case the zero row and zero column of the table contains only 0.

The only nontrivial part of the proof is left-distributivity.

VW 5. $\mathcal{R}$ and $\mathcal{S}$ are left-distributive: For a particular y and z in a system, to show that $x(y + z) = xy + xz$ for all x, look at the y column, the z column, and the $y + z$ column in the multiplication table. The vector sum of the y and z columns should equal the $y + z$ column. For example, in both $\mathcal{R}$ and $\mathcal{S}$, $(u + 1) + (u + 2) = 2u$. From the multiplication table for $\mathcal{R}$ these columns are as follows:

$$
\begin{pmatrix} u + 1 \\ 0 \\ u + 1 \\ 2u + 2 \\ u + 2 \\ 2 \\ 2u \\ 2u + 1 \\ u \\ 1 \end{pmatrix}
+
\begin{pmatrix} u + 2 \\ 0 \\ u + 2 \\ 2u + 1 \\ 2u + 2 \\ u \\ 2 \\ u + 1 \\ 1 \\ 2u \end{pmatrix}
=
\begin{pmatrix} 2u \\ 0 \\ 2u \\ u \\ 1 \\ u + 2 \\ 2u + 2 \\ 2 \\ u + 1 \\ 2u + 1 \end{pmatrix}.
$$

Left-distributivity follows by checking the remaining cases.   ■

Inspection of the multiplication table for $\mathcal{R}$ shows that $x^2 = 2$ for $x \neq 0, 1, 2$. Not as obvious is the fact that in $\mathcal{S}$, $x^2 = x + 1$ whenever $x \neq 0, 1, 2$. Verification of this is left as an exercise. $\mathcal{R}$ and $\mathcal{S}$ are not isomorphic, since any isomorphism f from $\mathscr{R}$ to $\mathscr{S}$ would have to fix 0, 1, and 2. Therefore for any $y \in \mathscr{S}$,

$$y^2 = (f(x))^2 = f(x^2) = f(2) = 2,$$

a contradiction.

The system $\mathscr{S}$ is not associative since

$$(2u + 1) \circ_{\mathscr{S}} [(u + 1) \circ_{\mathscr{S}} (u + 2)] = (2u + 1) \circ_{\mathscr{S}} (2u) = 1$$

but

$$[(2u + 1) \circ_{\mathscr{S}} (u + 1)] \circ_{\mathscr{S}} (u + 2) = 2u \circ_{\mathscr{S}} (u + 2) = 2u + 2.$$

It is left as an exercise to show that $\mathscr{R}$ is associative.

As is implied by the results in (Hall 1986, p. 214), given $(\mathscr{R}, +)$ the multiplication table for $\mathscr{R}$ can be constructed completely from the following information:

R1. $0, 1, 2$, commute with all elements of $\mathscr{R}$.
R2. $\mathscr{R}$ is left-distributive.
R3. $(a + bu)^2 = 2$ whenever $b \neq 0$.

As an example, consider $(u + 1)(2u + 1) = (u + 1)(u + 1 + u) = (u + 1)(u + 1) + (u + 1)u = 2 + (u + 1)(u + 1 + 2) = 2 + (u + 1)(u + 1) + (u + 1)2 = 2 + 2 + 2(u + 1) = 1 + 2u + 2 = 2u$.

Similarly the multiplication table for $\mathscr{S}$ can be constructed from R1, R2, and

S3. $(a + bu)^2 = (a + 1 + bu)$ whenever $b \neq 0$.

The chapter concludes with a description of a collineation (i.e., isomorphism) between the planes determined by $\mathscr{R}$ and $\mathscr{S}$. For tables of Hall's other two systems, and isomorphisms among all four planes, see (Fabbri 1993). See also (Stevenson 1972, p. 88) for the statement of the following theorem, which is proved by verifying all cases.

■ **THEOREM 16.** For all $x, y \in \{a + bu : a, b \in \mathbb{Z}_3\}$,

$$x \circ_{\mathscr{R}} y = (x + 2) \circ_{\mathscr{S}} y + y.$$

For example, $(u + 1) \circ_{\mathscr{R}} (u + 2) = u$, while

$$(u + 1 + 2) \circ_{\mathscr{S}} (u + 2) + (u + 2)$$
$$= u \circ_{\mathscr{S}} (u + 2) + (u + 2) = 1 + (u + 2) = u.$$

■ **THEOREM 17.** The projective planes over the nonisomorphic Veblen-Wedderburn systems $\mathscr{R}$ and $\mathscr{S}$ are isomorphic.

Proof: Define the mapping from the points of the plane over $\mathscr{R}$ to the points of the plane over $\mathscr{S}$ as

$$(p, q, 1) \rightarrow (p, q + 2, 1),$$

$$(p, 1, 0) \rightarrow (p, 1, 0),$$

$$(1, 0, 0) \rightarrow (1, 0, 0).$$

Clearly the map is one-one and onto. It will be shown that lines are mapped to lines as

$$[1, b, c] \rightarrow [1, b, b + c],$$

$$[0, 1, c] \rightarrow [0, 1, c + 1],$$

$$[0, 0, 1] \rightarrow [0, 0, 1].$$

CASE 1. Suppose that $(p, q, 1) \in [1, b, c]$ in the plane over $\mathscr{R}$. Then $p + q \circ_{\mathscr{R}} b + c = 0$. Thus $p + (q + 2) \circ_{\mathscr{S}} b + b + c = 0$, so $(p, q + 2, 1) \in [1, b, b + c]$ in the plane over $\mathscr{S}$.

CASE 2. Suppose that $(p, q, 1) \in [0, 1, c]$ in the plane over $\mathscr{R}$. Then $q + c = 0$, and therefore $(q + 2) + (c + 1) = 0$, so $(p, q + 2, 1) \in [0, 1, c + 1]$ in the plane over $\mathscr{S}$.

CASE 3. Suppose that $(p, q, 1) \in [0, 0, 1]$ in the plane over $\mathscr{R}$. Then $1 = 0$, a contradiction.

CASE 4. Suppose that $(p, 1, 0) \in [1, b, c]$ in the plane over $\mathscr{R}$. Then $p + b = 0$, so $p + b + 0 \circ_{\mathscr{S}} (b + c) = 0$, and $(p, 1, 0) \in [1, b, b + c]$ in the plane over $\mathscr{S}$.

CASE 5. Suppose that $(p, 1, 0) \in [0, 1, c]$ in the plane over $\mathscr{R}$. Then $1 = 0$, a contradiction.

CASE 6. Suppose that $(p, 1, 0) \in [0, 0, 1]$ in the plane over $\mathscr{R}$. Then $(p, 1, 0) \in [0, 0, 1]$ in the plane over $\mathscr{S}$.

CASE 7. Suppose that $(1, 0, 0) \in [1, b, c]$ in the plane over $\mathscr{R}$. Then $1 = 0$, a contradiction.

CASE 8. Suppose that $(1, 0, 0) \in [0, 1, c]$ in the plane over $\mathscr{R}$. Then $(1, 0, 0) \in [0, 1, c + 1]$ in the plane over $\mathscr{S}$.

CASE 9. Suppose that $(1, 0, 0) \in [0, 0, 1]$ in the plane over $\mathscr{R}$. Then $(1, 0, 0) \in [0, 0, 1]$ in the plane over $\mathscr{S}$.   ■

It can be shown that any collineation mapping the plane over $\mathscr{R}$ to itself takes the line at infinity $[0, 0, 1]$ to itself (see Room and Kirkpatrick (1973), sec. 4.3, for the proof). It is also known that given two lines ℓ and m in a projective plane $(\mathscr{P}', \mathscr{L}')$, the affine planes obtained by removing ℓ and m are isomorphic if and only if there is an isomorphism of the projective plane that maps ℓ to m. Thus removing ℓ_∞ from the projective plane over $\mathscr{R}$ results in a plane nonisomorphic to the one obtained by removing any other line.

In this chapter the examples have all been of coordinate planes, so the question arises as to whether non-Desarguesian affine and projective planes can be coordinatized. The answer is yes, but the coordinate systems obtained are not unique as was seen in the case of the plane coordinatized by $\mathscr{R}$ and $\mathscr{S}$ above. The systems obtained depend on the choice of the zero point and the line at infinity. Also there are several different methods of coordinatizing non-Desarguesian planes. In particular, see Hall (1986 p. 211) for a method of coordinatizing non-Desarguesian planes.

EXERCISES

The $\mathscr{R}$ and $\mathscr{S}$ referred to below are the systems whose tables are given in this section.

1. Verify that the $u + 1$ and the $u + 2$ columns of the multiplication table for $\mathscr{S}$ sum as vectors to the $2u$ column.

2. Verify that $x^2 = x + 1$ for all $x \neq 0, 1, 2$, in $\mathscr{S}$.

3. Using only $R1$, $R2$, and $R3$ compute

$$(2u + 1) \circ_{\mathscr{R}} (u + 1),$$

$$(2u + 2) \circ_{\mathscr{R}} u,$$

$$(u + 1) \circ_{\mathscr{R}} (u + 2).$$

4. Using only $R1$, $R2$, and $S3$, compute

$$(2u + 1) \circ_{\mathscr{S}} (u + 1),$$

$$(2u + 2) \circ_{\mathscr{S}} u,$$

$$(u + 1) \circ_{\mathscr{S}} (u + 2).$$

*5. (Hughes and Piper 1973, lem. 9.1, p. 85). Prove that the system $\mathscr{R}$ is associative.

6. Verify the statement of Theorem 16 in the following cases:

$$(2u + 1) \circ_{\mathscr{R}} (u + 1),$$

$$(2u + 2) \circ_{\mathscr{R}} u,$$

$$(u + 1) \circ_{\mathscr{R}} (u + 2).$$

SUGGESTED READING

Peter Dembowski, *Finite Geometries*, New York. Springer-Verlag, 1968.

Dana Fabbri, Non-Desarguesian projective planes of order nine and related topics, Senior honors thesis, University of Massachusetts, May 1993.

Marshall Hall, *Combinatorial Theory*, 2d, ed., New York. Wiley, 1986.

Marshall Hall, Projective planes, Trans. AMS **54** (1943), 229–277.

Daniel R. Hughes and Fred C. Piper, *Projective Planes*, Springer-Verlag Graduate Texts in Mathematics, New York. Springer-Verlag, 1973.

C. W. H. Lam, G. Kolesova, and L. Thiel, A Computer Search for Finite Projective Planes of Order 9, *Discrete Math.* **92** (1991), 187–195.

T. G. Room and P. B. Kirkpatrick, *Miniquaternion Geometry*, Cambridge. Cambridge University Press, 1971.

Frederick Stevenson, *Projective Planes*, San Francisco. Freeman, 1972. Reprint, Washington, NJ: Polygonal Press, 1992.

O. Veblen and J. H. Maclagan-Wedderburn, Non-Desarguesian and non-Pascalian geometries, *Trans. AMS* **8** (1907), 379–388.

CHAPTER 6

Affine Space

To define nonplanar affine spaces, in addition to the points and lines, a third set consisting of *planes* must be specified. Furthermore the definition of parallel lines must be modified to require such lines to be *coplanar* (i.e., lying in the same plane) as well as nonintersecting. Synthetic affine spaces are defined in the following section.

6.1 SYNTHETIC AFFINE SPACE

Here, after defining synthetic affine spaces, we present several examples of such spaces, together with examples showing the independence of the axioms.

Definition. An *affine space* $\mathscr{A}$ is an ordered triple $(\mathscr{P}, \mathscr{L}, \mathscr{E})$, where $\mathscr{P}$ is a nonempty set whose elements are called *points*, and $\mathscr{L}$ and $\mathscr{E}$ are nonempty collections of subsets of $\mathscr{P}$ called *lines* and *planes*, respectively, which satisfy the following axioms:

AS1. If P and Q are distinct points, there is a unique line ℓ such that $P \in \ell$ and $Q \in \ell$. (This line is denoted $\ell(P, Q)$).

AS2. If P, Q, and R are distinct noncollinear points, there is a unique plane (denoted $\{\overline{PQR}\}$) containing them.

AS3. (The parallel axiom). If P is a point not contained in a line ℓ, there is a unique line m such that $P \in m$ and m is parallel to ℓ. (m is *parallel* to ℓ, written $m \| \ell$, if $m \cap \ell = \varnothing$, and there is a plane α containing m and ℓ as subsets.)

AS4. (Transitivity of parallelism). If ℓ, m, and k are distinct lines with $\ell \| m$ and $m \| k$, then $\ell \| k$.

In view of *AS4*, the relation "is parallel or equal to" is an equivalence relation on the lines of $\mathscr{A}$.

□**EXAMPLE 1:** Let $(\mathscr{P}, \mathscr{L})$ be any affine plane with $\mathscr{E} = \{\mathscr{P}\}$. Then Axioms *AS1* and *AS3* hold, since they are true in any affine plane; $\mathscr{P}$ is the unique plane containing any three noncollinear points, so *AS2* holds; and the relation "is parallel or equal to" is transitive, since this is the case in affine planes, so *AS4* holds as well. Note that parallel lines here are simply nonintersecting lines, since all lines are coplanar. □

□**EXAMPLE 2: (Coordinate affine spaces).** Let F be any field. The three-dimensional vector space F^3 is an affine space (denoted $AG(F^3)$), where

$$\mathscr{P} = F^3 = \left\{ \mathbf{x} = (x_1, x_2, x_3) : x_i \in F \right\},$$

and a subset $\ell \subseteq \mathscr{P}$ is a line when, for some $\mathbf{u}$ and $\mathbf{v}$ in F^3, with $\mathbf{v} \neq \mathbf{0}$,

$$\ell = \left\{ \mathbf{u} + t\mathbf{v} : t \in F \right\}.$$

A subset α of $\mathscr{P}$ is in $\mathscr{E}$ if and only if it has the form

$$\alpha = \left\{ \mathbf{u} + t\mathbf{v} + s\mathbf{w} : t, s \in F \right\},$$

where $\mathbf{v}$ and $\mathbf{w}$ are nonzero vectors, neither of which is a multiple of the other. □

More generally, any F^n or $\mathbb{D}^n$ for $\mathbb{D}$ a division ring is an affine space (called the *n-dimensional affine space over F (or $\mathbb{D}$)* and written $AG(F^n)$ or $AG(\mathbb{D}^n)$), where the line containing the vectors $\mathbf{x}$ and $\mathbf{y}$ is

$$\ell(\mathbf{x}, \mathbf{y}) = \left\{ \mathbf{x} + t(\mathbf{y} - \mathbf{x}) : t \in \mathbb{D} \right\} = \left\{ t\mathbf{y} + (1 - t)\mathbf{x} : t \in \mathbb{D} \right\},$$

and the distinct lines $\ell(\mathbf{x}, \mathbf{y})$ and $\ell(\mathbf{z}, \mathbf{w})$ are parallel exactly when $\mathbf{y} - \mathbf{x}$ is a nonzero multiple of $\mathbf{z} - \mathbf{w}$. The plane containing the noncollinear vectors $\mathbf{x}$, $\mathbf{y}$, and $\mathbf{z}$ is

$$\left\{ \overline{\mathbf{x}, \mathbf{y}, \mathbf{z}} \right\} = \left\{ \mathbf{x} + t(\mathbf{y} - \mathbf{x}) + s(\mathbf{z} - \mathbf{x}) : t, s \in \mathbb{D} \right\}.$$

Clearly $\mathbf{x}$ and $\mathbf{y}$ are in $\ell(\mathbf{x}, \mathbf{y})$ as described above. If they are also in $\ell(\mathbf{z}, \mathbf{w})$, then $\mathbf{x} = \mathbf{z} + t(\mathbf{w} - \mathbf{z})$ and $\mathbf{y} = \mathbf{z} + s(\mathbf{w} - \mathbf{z})$. Thus $\mathbf{x} - \mathbf{y} = (t - s)(\mathbf{w} - \mathbf{z})$, and any point of the form $\mathbf{x} + r(\mathbf{y} - \mathbf{x})$ can be written as $\mathbf{z} + (t + r(s - t))(\mathbf{w} - \mathbf{z})$. Thus

$$\left\{ \mathbf{x} + t(\mathbf{y} - \mathbf{x}) : t \in \mathbb{D} \right\} \subseteq \left\{ \mathbf{z} + s(\mathbf{w} - \mathbf{z}) : s \in \mathbb{D} \right\}.$$

A similar argument shows that the reverse inclusion holds as well, so $AS1$ is verified.

If $\mathbf{x}$, $\mathbf{y}$ and $\mathbf{z}$ are noncollinear vectors, they are evidently in the plane $\{\mathbf{x}, \mathbf{y}, \mathbf{z}\}$ described above. If they are also in $\{\mathbf{u}, \mathbf{v}, \mathbf{w}\}$, then $\mathbf{x} = \mathbf{u} + t(\mathbf{v} - \mathbf{u}) + s(\mathbf{w} - \mathbf{u})$, and similarly for $\mathbf{y}$ and $\mathbf{z}$. Thus $\mathbf{y} - \mathbf{x}$ and $\mathbf{z} - \mathbf{x}$ are of the form $r(\mathbf{v} - \mathbf{u}) + s(\mathbf{w} - \mathbf{u})$, and

$$\{\mathbf{x}, \mathbf{y}, \mathbf{z}\} \subseteq \{\mathbf{u}, \mathbf{v}, \mathbf{w}\}.$$

A similar argument shows the reverse inclusion, so $AS2$ holds.

If $\mathbf{y} - \mathbf{x} = r(\mathbf{z} - \mathbf{w})$ for some nonzero r, then any element in $\ell(\mathbf{x}, \mathbf{y})$ is of the form $\mathbf{x} + t(\mathbf{z} - \mathbf{w})$. If the lines intersect at a vector $\mathbf{v}$, then $\mathbf{v} = \mathbf{x} + t(\mathbf{z} - \mathbf{w}) = \mathbf{w} + s(\mathbf{z} - \mathbf{w})$. Thus $\mathbf{x} = \mathbf{w} + (s - t)(\mathbf{z} - \mathbf{w}) \in \ell(\mathbf{z}, \mathbf{w})$. Similarly $\mathbf{y} \in \ell(\mathbf{z}, \mathbf{w})$, so the lines are the same. Thus if $\mathbf{y} - \mathbf{x}$ is a nonzero multiple of $\mathbf{z} - \mathbf{w}$, $\ell(\mathbf{x}, \mathbf{y})$, and $\ell(\mathbf{z}, \mathbf{w})$ are either equal or have empty intersection. In the latter case, they are parallel, since $\mathbf{y} = \mathbf{x} + r(\mathbf{z} - \mathbf{w}) = \mathbf{x} + r(\mathbf{z} - \mathbf{x}) - r(\mathbf{w} - \mathbf{x})$, so $\mathbf{y}$ is in the plane containing $\mathbf{x}$, $\mathbf{z}$, and $\mathbf{w}$.

Conversely, if $\ell(\mathbf{x}, \mathbf{y}) \| \ell(\mathbf{z}, \mathbf{w})$, $\mathbf{y}$ is coplanar with the other three points; hence $\mathbf{y} = \mathbf{x} + t(\mathbf{z} - \mathbf{x}) + s(\mathbf{w} - \mathbf{x})$. If t is different from $-s$, it follows that

$$\mathbf{y} + (t + s - 1)\mathbf{x} = t\mathbf{z} + s\mathbf{w}.$$

Multiplying through by $(t + s)^{-1}$ gives

$$\mathbf{x} + (t + s)^{-1}(\mathbf{y} - \mathbf{x}) = \mathbf{z} + (t + s)^{-1}s(\mathbf{w} - \mathbf{z}),$$

so the lines intersect, a contradiction. Thus $t + s = 0$ and $\mathbf{y} = \mathbf{x} + t(\mathbf{z} - \mathbf{x}) - t(\mathbf{w} - \mathbf{x})$, so $\mathbf{y} - \mathbf{x} = t(\mathbf{z} - \mathbf{w})$.

Thus, if $\mathbf{z}$ is not in $\ell(\mathbf{x}, \mathbf{y})$, the unique line containing $\mathbf{z}$ and parallel to $\ell(\mathbf{x}, \mathbf{y})$ is

$$\ell(\mathbf{z}, \mathbf{z} - \mathbf{y} + \mathbf{x}) = \{\mathbf{z} + t(\mathbf{y} - \mathbf{x}) : t \in \mathbb{D}\}.$$

The transitivity of parallelism is a direct result of the characterization of parallel lines given above.

A third example is the smallest affine space that contains more than one plane.

☐ **EXAMPLE 3:** Let $\mathscr{P} = \{A, B, C, D, E, F, G, H\}$, and let $\mathscr{L}$ be the 28 two-element subsets of $\mathscr{P}$. $\mathscr{E}$ consists of the subsets

{ABEF}	{CDGH}	{BHCE}	{AGDF}	{ABGH}
{ACEG}	{CDEF}	{BHDF}	{GHEF}	{ABCD}
{ACFH}	{BDEG}	{ADEH}	{BCGF}	

It is left as an exercise to show that this space is isomorphic to $AG(\mathbb{Z}_2^3)$.
☐

Independence

Showing the independence of the axioms for affine space requires four models.

☐**EXAMPLE 4:** Let $\mathscr{P} = \{A, B, C\}$; $\mathscr{L} = \{\{ABC\}, \{AC\}, \{B\}\}$; $\mathscr{E} = \{\{ABC\}\}$. Axiom $AS1$ clearly fails here. Since all points are collinear, Axiom $AS2$ holds vacuously, as does Axiom $AS4$ because only the lines $\{B\}$ and $\{AC\}$ are parallel. Axiom $AS3$ is satisfied, since the only possibilities for a point not being on a line are $B \notin \{AC\}$ and $A, C \notin \{B\}$. Thus Axiom $AS1$ is independent of the other axioms.   ☐

☐**EXAMPLE 5:** Let $(\mathscr{P}, \mathscr{L})$ be the affine plane of order 2 with points A, B, C, and D; let $\mathscr{E} = \{\{ABC\}, \{ABCD\}\}$. Axiom $AS2$ fails since the noncollinear points A, B, and C are in two planes. Axioms $AS1$, $AS3$, and $AS4$ hold as they did in $(\mathscr{P}, \mathscr{L})$ (because all lines are in plane $\{ABCD\}$, parallel lines are simply nonintersecting lines).   ☐

☐**EXAMPLE 6:   (The Fano plane).** Let $(\mathscr{P}, \mathscr{L})$ be the seven-point, seven-line projective plane (given in Example 5 of Chapter 2), and let $\mathscr{E} = \{\mathscr{P}\}$. Here Axiom $AS3$ fails, but $AS1$, $AS2$, and $AS4$ hold, with $AS4$ holding vacuously.   ☐

☐**EXAMPLE 7:** The relevant example to prove the independence of $AS4$, first given in (Sasaki 1952), is more complicated. The "points" are the sixteen 4-tuples (x, y, z, w) in $(\mathbb{Z}_2)^4$, while $\mathscr{L}$ is the set of all two-element subsets of $\mathscr{P}$ (there are 120 of them). We will define, for each triple of points, a fourth point, and these four-point sets are, by definition, the planes. First, take $\mathscr{E}$ to be the set of all points that have fourth coordinate equal to 0. Now define a mapping ϕ on $\mathscr{P}$ that maps each point to itself except that $\phi(1, 0, 0, 0) = (0, 1, 0, 0)$ and $\phi(0, 1, 0, 0) = (1, 0, 0, 0)$.

Let P, Q, and R be distinct points. If any of P, Q, and R are not in $\mathscr{E}$, then $\{\overline{PQR}\} = \{P, Q, R, P + Q + R\}$, the addition done componentwise in $\mathbb{Z}_2$. Thus the plane determined by $(0, 0, 1, 0)$, $(1, 0, 1, 0)$, and $(0, 0, 1, 1)$ contains as its fourth point $(1, 0, 1, 1)$. If P, Q, and R are all in $\mathscr{E}$, first look at $\phi(P), \phi(Q), \phi(R)$, and a fourth point, $\phi(P) + \phi(Q) + \phi(R)$, then take the image of each of these new points under ϕ (noting that ϕ^2 is the identity) to get the plane $\{P, Q, R, \phi(\phi(P) + \phi(Q) + \phi R))\}$ (ϕ does not necessarily preserve addition). For example, the fourth point on the plane determined by $(0, 0, 0, 0)$, $(0, 1, 0, 0)$, and $(1, 0, 1, 0)$ is $(0, 0, 1, 0)$.

The first three axioms for affine space hold in Sasaki's example, but if $P = (0, 0, 0, 0)$, $Q = (0, 1, 0, 0)$, $R = (0, 0, 1, 1)$, $S = (1, 0, 1, 1)$, $A = (0, 0, 1, 0)$, and $B = (1, 0, 1, 0)$, then $\ell(P, Q) \| \ell(A, B)$ and $\ell(A, B) \| \ell(R, S)$.

Yet P, Q. R, and S are not coplanar since $P + Q + R = (0, 1, 1, 1) \neq S$. Thus $\ell(P, Q)$ is not parallel to $\ell(R, S)$, and the transitivity of parallelism fails.   □

EXERCISES

1. a. Find a second point on the line containing $(2, 1, 3, 4)$ and parallel to $\ell((0, 0, 0, 0), (1, 2, 3, 4))$ in $AG(\mathbb{R}^4)$.

   b. List the points on $\ell((1, 3, 2, 2), (2, 1, 3, 0))$ in $AG(\mathbb{Z}_5^4)$.

2. Show that the plane affine space given in Example 3 is isomorphic to the affine space $AG(\mathbb{Z}_2^3)$.

3. List the points on the plane containing $(0, 1, 0)$, $(0, 2, 2)$, and $(1, 2, 1)$ in $AG(\mathbb{Z}_3^3)$.

4. The following questions pertain to $AG(\mathbb{Z}_7^4)$.

   a. Are $(2, 1, 3, 0)$, $(3, 1, 4, 0)$, and $(1, 3, 2, 5)$ collinear? Why or why not?

   b. Is $(3, 1, 2, 5)$ coplanar with $(0, 0, 0, 0)$, $(2, 1, 3, 4)$, and $(1, 1, 1, 1)$? Explain.

   c. Is $(4, 4, 4, 1)$ on the line containing $(3, 3, 2, 0)$ and parallel to $\ell((1, 1, 1, 3), (1, 1, 1, 2))$? Explain.

   d. How many points are on each line? How many points are in each plane?

5. Recall that the field $GF(4) \cong \mathbb{Z}_2[u]/(u^2 + u + 1)$. The following questions pertain to the affine space $(GF(4))^3$.

   a. Are $(1, u, 1)$, $(1, u, u + 1)$, and $(u + 1, 1, u)$ collinear?

   b. Are $(u, u + 1, 0)$, $(0, 1, u)$, $(0, 0, 0)$, and $(u, u, 1)$ coplanar?

   c. List the points on the line containing $(1, u + 1, 1)$ and parallel to $\ell((0, 1, u), (u, 1, u + 1))$.

6. Recall that $GF(9) \cong \mathbb{Z}_3[u]/(u^2 + 1)$. The following questions pertain to $(GF(9))^4$.

   a. Are the points $(1, u, 0, 1)$, $(2, 2u, 1, u)$, and $(u + 2, u, u, u)$ collinear?

   b. Are the points $(1, u, 1, 2)$, $(u + 1, u, 2, 2)$, $(u, u, 1, 1)$, and $(0, 0, 2u, u)$ coplanar?

   c. Describe the line containing $(1, u, u + 1, u)$ and parallel to $\ell((0, 0, 0, 0), (1, 0, u, 1))$.

7. Find the line of intersection of the planes $\overline{\{(0, 0, 0,), (1, 2, 1), (4, 3, 1)\}}$ and $\overline{\{(0, 0, 0), (2, 1, 1), (3, 1, 1)\}}$ in $AG(\mathbb{Z}_5^3)$.

8. Prove that $(\mathbb{Z}_4)^3$, with lines and planes defined as above, is not an affine space.

9. Prove that $\mathbf{x}$, $\mathbf{y}$, $\mathbf{z}$, and $\mathbf{x} - \mathbf{y} + \mathbf{z}$ are the vertices of a parallelogram in any $AG(F^n)$.

10 a. Prove that if two planes in $AG(F^3)$ intersect, their intersection is a line.

   b. Show, by giving an example of two planes whose intersection is a point, that the statement in (a) is no longer true in $AG(\mathbb{R}^4)$.

   c. Is the statement in (a) false in an arbitrary $AG(F^4)$?

11. Show that any line through the origin of F^n consists of the multiples of some nonzero vector, and conversely.

12. Show that any plane through the origin of F^n is of the form $\{s\mathbf{x} + t\mathbf{y}: s, t \in F\}$ where $\mathbf{x}$ is not collinear with both the origin and $\mathbf{y}$.

13. Prove that in any $AG(F^n)$, if two points are in a plane, the line containing them is a subset of that plane.

14. Suppose that $\mathbf{0}$, $\mathbf{x}$, and $\mathbf{y}$ are noncollinear vectors in F^n. Suppose that $\mathbf{w} \notin \overline{\{\mathbf{0}, \mathbf{x}, \mathbf{y}\}}$, and let a and b be elements of F. Prove that $\overline{\{\mathbf{0}, \mathbf{x}, \mathbf{y}\}} \cap \overline{\{\mathbf{0}, \mathbf{w} + a\mathbf{x}, \mathbf{w} + b\mathbf{y}\}} = \ell(\mathbf{0}, a\mathbf{x} - b\mathbf{y})$.

15. Suppose that $\mathbf{0}$, $\mathbf{x}$, and $\mathbf{y}$ are noncollinear points in F^n, with $\mathbf{z}$ on the plane they determine but not collinear with $\mathbf{0}$ and $\mathbf{y}$. Prove that there is a unique element a in F such that $\ell(\mathbf{0}, \mathbf{z}) = \ell(\mathbf{0}, \mathbf{x} + a\mathbf{y})$.

16. Show that in any affine space over $\mathbb{Z}_2$, if $\mathbf{x}$, $\mathbf{y}$, and $\mathbf{z}$ are noncollinear, the plane they determine is the four-point set $\{\mathbf{x}, \mathbf{y}, \mathbf{z}, \mathbf{x} + \mathbf{y} + \mathbf{z}\}$.

*17. Show that $(\mathbb{Z}_k)^n$ is not an affine space unless k is prime.

*18. If $\mathscr{R}$ is the Veblen-Wedderburn system whose multiplication table is given in Chapter 5, show that $\mathscr{R}^3$, with lines and planes defined as for $AG(F^3)$, is not an affine space.

6.2 FLATS IN AFFINE SPACE

In general, the subobjects of a mathematical object are the subsets that share its properties. For example, the (linear) subspaces of the vector spaces F^3 are the subsets of F^3 that are closed under addition and scalar multiplication; each subspace is a vector space in its own right.

Thus the "subspaces" (usually called *flats*) of an affine space $\mathscr{A}$ are the subsets of $\mathscr{P}$ that are themselves affine spaces whose lines are lines in $\mathscr{L}$. They are defined as follows.

Definition. A subset $\mathscr{F}$ of the points of an affine space $\mathscr{A} = (\mathscr{P}, \mathscr{L}, \mathscr{E})$ is a *flat* of $\mathscr{A}$ if whenever P and Q are distinct points in $\mathscr{F}$, $\ell(P, Q) \subseteq \mathscr{F}$. (When the context is clear, we will refer to $\mathscr{F}$ as being a flat of $\mathscr{P}$.)

It will follow from Theorem 14 that the flats of $AG(\mathbb{D}^n)$ are the translates of the linear subspaces of $\mathbb{D}^n$. See also Exercise 7 below.

■ **THEOREM 1.** If $(\mathscr{P}, \mathscr{L}, \mathscr{E})$ is an affine space, then each line in $\mathscr{L}$ and each plane in $\mathscr{E}$ is a flat of $\mathscr{P}$.

Proof: If ℓ is a line with P and Q in ℓ, then $\ell = \ell(P, Q)$, and ℓ is clearly a flat.

If α is a plane containing P and Q, and R is a point in $\ell(P, Q)$, take S in α but not in $\ell(P, Q)$. (If no such S exists, then $\alpha = \ell(P, Q)$, which has been shown to be a flat.) Through S there is a unique line m parallel to $\ell(P, Q)$, which implies that S and $\ell(P, Q)$ are in some plane. But the only plane containing P, Q, and S is α, so $\ell(P, Q) \subseteq \alpha$. ■

If $\{\mathcal{F}_i\}_{i \in I}$ is a collection of flats of an affine space, then for any points P and Q in $\cap_{i \in I} \mathcal{F}_i$, $\ell(P, Q)$ is a subset of $\mathcal{F}_i$ for every i, and thus is a subset of $\cap_{i \in I} \mathcal{F}_i$ as well. Therefore arbitrary intersections of flats are always flats.

In view of the remarks above, it follows that for any *subset* $\mathcal{T}$ of $\mathcal{P}$, the smallest *flat* of $\mathcal{P}$ containing $\mathcal{T}$ exists and is given by

$$\overline{\mathcal{T}} = \cap \{\mathcal{F} : \mathcal{F} \text{ a flat of } \mathcal{P} \text{ and } \mathcal{T} \subseteq \mathcal{F}\}.$$

Clearly if $\mathcal{T}$ is already a flat of $\mathcal{P}$, then $\overline{\mathcal{T}} = \mathcal{T}$. The basic properties of the mapping taking $\mathcal{T}$ to $\overline{\mathcal{T}}$ are summarized below.

■ **THEOREM 2.** For $\mathcal{S}$ and $\mathcal{T}$ subsets of $\mathcal{P}$,

   i. $\mathcal{S} \subseteq \overline{\mathcal{S}}$,
   ii. $\overline{\overline{\mathcal{S}}} = \overline{\mathcal{S}}$,
   iii. If $\mathcal{S} \subseteq \mathcal{T}$, then $\overline{\mathcal{S}} \subseteq \overline{\mathcal{T}}$,
   vi. If $\mathcal{S} \subseteq \overline{\mathcal{T}}$, then $\overline{\mathcal{S}} \subseteq \overline{\mathcal{T}}$.

Proof: Part i is trivially true. To show part ii note that containment one way ($\supseteq$) follows from (i). The reverse inclusion follows from the fact that $\overline{\mathcal{S}}$ is one of the flats being intersected to form $\overline{\overline{\mathcal{S}}}$. If $\mathcal{S} \subseteq \mathcal{T}$, and $\mathcal{T}$ is contained in any flat, $\mathcal{S}$ is a subset of that flat as well; thus the flats whose intersection is $\overline{\mathcal{S}}$ include those whose intersection is $\overline{\mathcal{T}}$, and $\overline{\mathcal{S}} \subseteq \overline{\mathcal{T}}$. Finally, if $\mathcal{S}$ is a subset of $\overline{\mathcal{T}}$, then $\overline{\mathcal{T}}$ is one of the sets whose intersection forms $\overline{\mathcal{S}}$, so $\overline{\mathcal{T}}$ contains $\overline{\mathcal{S}}$. ■

A mapping such as $\mathcal{S} \to \overline{\mathcal{S}}$ given above is an example of a *closure operator* on $\mathcal{P}$, defined below.

Definition. Let $\mathcal{X}$ be any set, and let $c : \mathbf{P}(\mathcal{X}) \to \mathbf{P}(\mathcal{X})$ be a function defined on the subsets of $\mathcal{X}$. Then c is called a *closure operator* on $\mathcal{X}$ if the following conditions hold for all subsets $\mathcal{M}$ and $\mathcal{N}$ of $\mathcal{X}$:

   1. $\mathcal{M} \subseteq c(\mathcal{M})$ (c is *increasing*).
   2. $c(c(\mathcal{M})) = c(\mathcal{M})$ (c is *idempotent*).
   3. If $\mathcal{M} \subseteq \mathcal{N}$, then $c(\mathcal{M}) \subseteq c(\mathcal{N})$ (c is *isotone*).

The pair $(\mathcal{X}, c)$ is called a *closure space* (or *closure system*), and a subset $\mathcal{M}$ of $\mathcal{X}$ is said to be *closed* if $\mathcal{M} = c(\mathcal{M})$.

Since the conditions above are exactly parts i–iii of Theorem 2, it follows that $(\mathscr{P}, \overline{})$ is a closure space. The set $\overline{\mathscr{S}}$ is called the *affine closure* or *affine hull* of $\mathscr{S}$.

For $\mathscr{F}$ a flat in an affine space $(\mathscr{P}, \mathscr{L}, \mathscr{E})$, define $\mathscr{L}_{\mathscr{F}}$ to be the set of lines contained in $\mathscr{F}$, namely, $\mathscr{L}_{\mathscr{F}} = \mathscr{L} \cap \mathbf{P}(\mathscr{F})$, where $\mathbf{P}(\mathscr{F})$ is the collection of all subsets of $\mathscr{F}$; similarly define $\mathscr{E}_{\mathscr{F}}$ to be the planes which are subsets of $\mathscr{F}$, namely $\mathscr{E}_{\mathscr{F}} = \mathscr{E} \cap \mathbf{P}(\mathscr{F})$. The points, lines, and planes in a flat form an affine space as can be seen in the following theorem.

■   **THEOREM 3.** Let $(\mathscr{P}, \mathscr{L}, \mathscr{E})$ be any nontrivial affine space that has at least three points on each line, and let $\mathscr{F}$ be a flat. Then $(\mathscr{F}, \mathscr{L}_{\mathscr{F}}, \mathscr{E}_{\mathscr{F}})$ is an affine space as well.

Proof: Axiom $AS1$ holds, since for P and Q in $\mathscr{F}$ the unique line they determine in $\mathscr{L}$ is a subset of $\mathscr{F}$. If P, Q, and R are noncollinear points, then $\{PQR\} \subseteq \mathscr{F}$ implies by part iv of Theorem 2 that $\{\overline{PQR}\} \subseteq \mathscr{F}$; hence Axiom $AS2$ holds. The parallel axiom follows from the fact that if $P \notin \ell$ with P and ℓ in $\mathscr{F}$, the line containing P and parallel to ℓ is in $\mathscr{F}$ as well. (See Exercise 5 below.) Finally, lines contained in $\mathscr{F}$ and parallel in $\mathscr{P}$ are also parallel in $\mathscr{F}$, from which the transitivity of parallelism follows.

If an affine space has exactly two points on each line, then *every* subset of its points is technically a flat. To make the preceding theorem hold in this case, the definition of flat is modified by adding the condition that with any three noncollinear points in $\mathscr{F}$, the plane they determine is a subset of $\mathscr{F}$. From now on, when we speak of a flat, we will mean to include this condition whenever each line has exactly two points.

EXERCISES

1. Show that $(\alpha, \mathscr{L}_{\alpha})$ is an affine plane whenever $(\mathscr{P}, \mathscr{L}, \mathscr{E})$ is an affine space and α is a plane in $\mathscr{E}$ whose points are not all collinear. (Note that this amounts to showing that planes in affine spaces are affine planes.)

2. Show that each line of any affine space contains the same number of points.

3. Suppose that $(\mathscr{P}, \mathscr{L}, \mathscr{E})$ is the affine space $AG(F^3)$ where $F \cong GF(p^n)$. Find the number of points, lines, and planes in $(\mathscr{P}, \mathscr{L}, \mathscr{E})$. Find the number of points on each line, the number of lines through each point, and the number of points in each plane.

4. Let $\mathscr{X}$ be a set, and let $c : \mathbf{P}(\mathscr{X}) \to \mathbf{P}(\mathscr{X})$ be a function. Prove that c is a closure operator if and only if the following condition holds for arbitrary subsets $\mathscr{M}$ and $\mathscr{N}$ of $\mathscr{X}$

$$\mathscr{M} \subseteq c(\mathscr{N}) \quad \text{if and only if} \quad c(\mathscr{M}) \subseteq c(\mathscr{N}).$$

5. Suppose that each line in an affine space $(\mathscr{P}, \mathscr{L}, \mathscr{E})$ has at least three points. Suppose that $\mathscr{F}$ is a flat of $\mathscr{P}$, $P \notin \ell(R, T)$ and $\ell(P, Q) \| \ell(R, T)$. Prove, using only the definition of a flat (as containing, with any two points, their line), that if P, R, and T are in $\mathscr{F}$, then $Q \in \mathscr{F}$.

6. Let $\mathscr{X}$ be a set, and let $\mathscr{J}$ be a collection of subsets of $\mathscr{X}$ such that $\mathscr{J}$ contains $\mathscr{X}$, and the intersection of any collection of sets in $\mathscr{J}$ is again in $\mathscr{J}$. Define $c : \mathbf{P}(\mathscr{X}) \to \mathbf{P}(\mathscr{X})$ by $c(\mathscr{M}) = \cap \{\mathscr{N} \in \mathscr{J} : \mathscr{M} \subseteq \mathscr{N}\}$. Prove that c is a closure operator.

*7. a. Recall that a subset $\mathscr{S} \subseteq F^n$ is a *linear subspace* if whenever $\mathbf{x}$ and $\mathbf{y}$ are in $\mathscr{S}$ and $t \in F$, then $\mathbf{x} + \mathbf{y}$ and $t\mathbf{x}$ are in $\mathscr{S}$. If $\mathscr{S}$ is a subspace of F^n and $\mathbf{w}$ a fixed element of F^n, prove that $\mathscr{S}_{\mathbf{w}} = \{\mathbf{w} + \mathbf{x} : \mathbf{x} \in \mathscr{S}\}$ is a flat in F^n.

   b. Conversely prove that for $\mathscr{F}$ a flat in $AG(F^n)$ and $\mathbf{u}$ any vector in $\mathscr{F}$, then $\{\mathbf{x} - \mathbf{u} : \mathbf{x} \in \mathscr{F}\}$ is a linear subspace of F^n.

6.3 DESARGUES' THEOREM

It is surprising, in view of the non-Desarguesian Moulton plane discussed in Chapter 3, that Desargues' Theorem can be *proved* in a space that has more than one plane. Once this is done, Theorem 3 can be used to show that the lines in a nonplanar affine space can be made into division rings, as in Chapter 3, and planes in a nonplanar affine space can be coordinatized as in Chapter 4.

■ **THEOREM 4. Desargues' Theorem for affine space.** Let $(\mathscr{P}, \mathscr{L}, \mathscr{E})$ be an affine space with more than one plane. Suppose that A, B, and C are noncollinear points such that $\ell(A, A') \| \ell(B, B') \| \ell(C, C')$ or $\ell(A, A') \cap \ell(B, B') \cap \ell(C, C') = P$. If $\ell(A, B) \| \ell(A', B')$ and $\ell(A, C) \| \ell(A', C')$, then $\ell(B, C) \| \ell(B', C')$.

Proof:

CASE 1. $\{\overline{ABC}\} \neq \{\overline{A'B'C'}\}$. If $\ell(B, B') \| \ell(C, C')$, the points B, B', C, and C' are coplanar. If the lines intersect at P, the four points are in the plane $\{\overline{PBC}\}$. Thus $\ell(B, C)$ and $\ell(B', C')$ are either parallel or they intersect. If they intersect, say, at point Q, then let $\ell(Q, D)$ be the line through Q parallel to $\ell(A, B)$ and $\ell(Q, E)$ be the line through Q parallel to $\ell(A, C)$ (see Fig. 6.1). Now $\{\overline{ABC}\}$ contains the noncollinear points Q, D, and E, whence $\{\overline{ABC}\} = \{\overline{QED}\}$. But by the transitivity of parallelism ($AS4$), $\ell(Q, D) \| \ell(A', B')$ and $\ell(Q, E) \| \ell(A'C')$; thus $\{\overline{A'B'C'}\}$ contains Q, D, and E, and $\{\overline{A'B'C'}\} = \{\overline{QED}\} = \{\overline{ABC}\}$, a contradiction.

CASE 2. A, B, C, A', B', and C' are coplanar. Let X be a point not coplanar with the six points above. Let m be the line containing X and parallel to $\ell(A, A')$ if $\ell(A, A') \| \ell(B, B')$; let $m = \ell(X, P)$ if $\ell(A, A') \cap \ell(B, B') = P$. Let X' be the point of intersection of m with the line

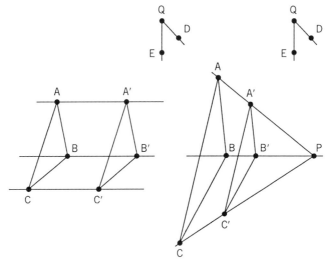

Figure 6.1

through A parallel to $\angle(A, X)$ (see Fig. 6.2). Now use Case 1 on triangles ABX and A'B'X' to get $\angle(B, X) \| \angle(B', X')$; use Case 1 on triangles ACX and A'C'X' to get $\angle(C, X) \| \angle(C', X')$. Finally, use Case 1 on triangles BCX and B'C'X' to get $\angle(B, C) \| \angle(B', C')$. ∎

■ **THEOREM 5.** Suppose that $(\mathscr{P}, \mathscr{L}, \mathscr{E})$ is an affine space with $\{\overline{OAB}\}$, $\{\overline{OAC}\}$, $\{\overline{OCD}\}$, and $\{\overline{OBD}\}$ distinct planes in $\mathscr{E}$. Then, if $\angle(O, E) \subseteq \{\overline{OAB}\} \cap \{\overline{OCD}\}$, there is an $F \in \mathscr{P}$ with $\angle(O, F) \subseteq \{\overline{OAC}\} \cap \{\overline{OBD}\}$.

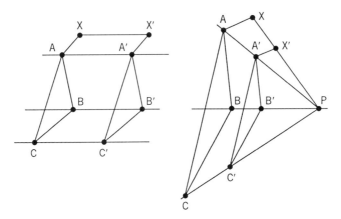

Figure 6.2

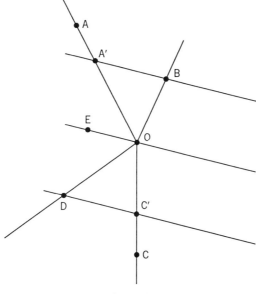

Figure 6.3

Proof: Let m be the line containing B parallel or equal to $\ell(O, E)$, and $m \cap \ell(O, A) = A'$ (see Fig. 6.3). Similarly there is a $C' \in \ell(O, C)$ with $\ell(C', D)$ parallel or equal to $\ell(O, E)$. Now $\ell(A', B) \| \ell(C', D)$ so that A', B, C', and D are coplanar. If $\ell(A', C') \cap \ell(B, D) = F$, then $F \in \ell(A', C') \subseteq \{\overline{OAC}\}$. But $F \in \ell(B, D) \subseteq \{\overline{OBD}\}$; hence $\ell(O, F)$ is in the intersection of the two planes. If $\ell(A', C') \| \ell(B, D)$, let k be the line containing O and parallel to $\ell(A', C')$. For any point $F \neq O$ in k, $\ell(O, F) \subseteq \{\overline{OAC}\} \cap \{\overline{OBD}\}$. ∎

EXERCISES

In the following exercises assume that each affine space has more than one plane so that Desargues' Theorem holds.

1. Prove that the converse of Desargues' Theorem holds in any affine space. [*Hint:* See Theorem 9 of Chapter 3.]

2. Prove Desargues' Theorem analytically for $AG(\mathbb{D}^n)$, where $\mathbb{D}$ is any division ring and $n \geq 2$.

3. Let $(\mathscr{P}, \mathscr{L}, \mathscr{E})$ be an affine space. Let A, B, C, D, E, and F be distinct points, with A, B, and C noncollinear, such that $\ell(A, B) \| \ell(D, E)$ and $\ell(A, C) \| \ell(D, F)$. Prove that $\{\overline{ABC}\} = \{\overline{DEF}\}$ or $\{\overline{ABC}\} \cap \{\overline{DEF}\} = \varnothing$.

*4. Let $(\mathscr{P}, \mathscr{L}, \mathscr{E})$ be an affine space with P, A, and B noncollinear. Let E, F, and G be noncollinear with $\ell(P, A) \| \ell(E, F)$ and $\ell(P, B) \| \ell(E, G)$. Let $D \in \{\overline{PAB}\}$, and let $\ell(E, H) \| \ell(P, D)$. Prove that H is in $\{\overline{EFG}\}$. [*Hint:* Try to get

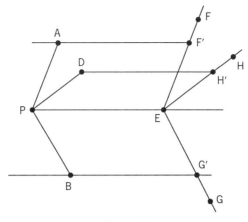

Figure 6.4

F', G', H', as shown in Fig. 6.4, and use Desargues' Theorem to show that $\ell(A, B)$ and $\ell(P, D)$ are either parallel or intersecting lines, which makes them coplanar. What about $\ell(F', G')$ and $\ell(E, H)$?]

6.4 COORDINATIZATION OF AFFINE SPACE

For any vector space $\mathcal{V}$ over a division ring $\mathbb{D}$, the affine space $AG(\mathcal{V})$ can be defined as in Example 2 above. Conversely, if $\mathcal{A} = (\mathcal{P}, \mathcal{L}, \mathcal{E})$ is any affine space in which Desargues' Theorem holds, it can be coordinatized by a division ring.

More precisely, it will be shown that if $(\mathcal{P}, \mathcal{L}, \mathcal{E})$ is either a non-planar affine space or a Desarguesian affine plane, it can be *coordinatized* so that $\mathcal{P} = \mathcal{V}$, $\ell \in \mathcal{L}$ if and only if $\ell = \{\mathbf{x} + t(\mathbf{y} - \mathbf{x}) : t \in \mathbb{D}\}$, and $\alpha \in \mathcal{E}$ if and only if $\alpha = \{\mathbf{x} + t(\mathbf{y} - \mathbf{x}) + s(\mathbf{z} - \mathbf{x}) : s, t \in \mathbb{D}\}$, where $\mathcal{V}$ is a vector space over the division ring $\mathbb{D}$.

It was proved in Theorem 4 that Desargues' Theorem holds in any affine space of dimension at least 3, that is, in any space that contains more than one plane. We will use the term *Desarguesian affine space* to mean either a Desarguesian affine plane as defined in Chapter 3 or an affine space in which not all points are coplanar.

■ **THEOREM 6.** In any Desarguesian affine space $(\mathcal{P}, \mathcal{L}, \mathcal{E})$, on each line ℓ in $\mathcal{L}$, relative to a specific O, and I, an addition and a multiplication can be defined under which ℓ is a division ring, and any two such rings constructed in the same space are isomorphic.

Proof: This proof follows almost immediately from the results in Sections 3.2, 3.4, 3.6, and 3.7 (see in particular Theorem 20). Any line in $\mathcal{L}$

is in some plane; each plane is Desarguesian; therefore each line ℓ can be made into a division ring that can be designated (ℓ, O, I). If (m, O', I') is another ring, then the lines ℓ and $k = \ell(O, I')$ (if $O = I'$, use $k = \ell(I, I')$ instead) are coplanar; hence the rings (ℓ, O, I) and (k, O, I') are isomorphic. Similarly (k, O, I') and (m, O', I') are isomorphic, so any pair of lines are isomorphic as division rings.   ■

Let $\mathbb{D}$ be any division ring isomorphic to all the lines in $(\mathscr{P}, \mathscr{L}, \mathscr{E})$ ($\mathbb{D}$ can be one of the lines, or an abstract division ring such as $\mathbb{Z}_p$, $GF(p^n)$ or $\mathbb{H}$). It will be shown that $\mathscr{P}$ is a vector space over $\mathbb{D}$. But, to do this, vector addition of points in $\mathscr{P}$, and scalar multiplication of elements in $\mathbb{D}$ by points in $\mathscr{P}$ must be defined, and for this an *origin* and *unit* vectors are selected.

First, select and fix any point O in $\mathscr{P}$. Let ℓ be any line containing O, and let I be any point on ℓ different from O. Now (ℓ, O, I) is a division ring with identities O and I. If $O \in m$, and I_m is any point of m other than O, the rings (ℓ, O, I) and (m, O, I_m) are isomorphic under the mapping $\phi_{\ell m} : \ell \to m$ defined by $\phi(O) = O, \phi(I) = I_m$, and for A in ℓ different from O and I, $\phi(A) (= A')$ is the point of m such that $(I, I_m) \| \ell(A, \phi(A))$ as shown in Fig. 6.5.

Select an abstract division ring $\mathbb{D}$ that is isomorphic to ℓ, and denote the isomorphism $\phi_\ell : \mathbb{D} \to \ell$. Let 0 and 1 be the identities of $\mathbb{D}$, and use lowercase letters at the beginning of the alphabet for its other elements. Thus $\phi_\ell(0) = O$ and $\phi_\ell(1) = I$. If m is any line containing O, denote by ϕ_m the composite $\phi_{\ell m} \circ \phi_\ell : \mathbb{D} \to m$. Thus for each ring element a and line k containing O, $\phi_k(a) = A$ for a unique point A in k. Conversely for any point B in $\mathscr{P}$, there is a unique ring element b such that $\phi_{\ell(O, B)}(b) = B$. This is the element such that $\ell(I, I_{\ell(O, B)}) \| \ell(\phi_\ell(b), B)$, as shown in

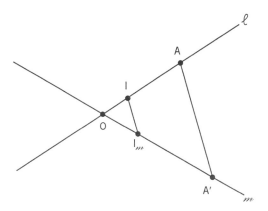

Figure 6.5

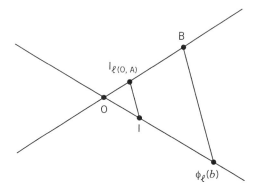

Figure 6.6

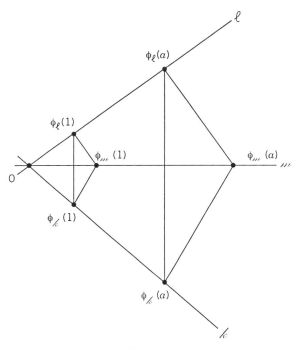

Figure 6.7

Fig. 6.6. It follows from Desargues' Theorem that

$$\ell\big(\phi_m(a),\,\phi_k(a)\big)\|\ell\big(\phi_m(1),\,\phi_m(1)\big)$$

whenever m and k are lines that intersect at O as shown in Fig. 6.7.
Vector addition and scalar multiplication can now be defined.

Definition. Let P and Q be points in the Desarguesian affine space
$(\mathcal{P},\mathcal{L},\mathcal{E})$. If O, P, and Q are on some line m (in particular, if P = Q or

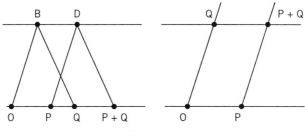

Figure 6.8

one of the points is O), then P + Q is the point defined as the sum of P and Q on (m, O, I_m), as illustrated in Fig. 6.8. If O, P, and Q are noncollinear, then P + Q is defined to be the unique point such that $\ell(O, P) \| \ell(Q, P + Q)$ and $\ell(O, Q) \| \ell(P, P + Q)$, whereby (O, P, P + Q, Q) is a parallelogram as shown in Fig. 6.8.

For O, P, and Q collinear points, Desargues' Theorem can be used to show that the definition of P + Q is independent of the choice of the auxiliary point B. (See Exercise 3a.)

■ **THEOREM 7.** Relative to any choice of O, the addition of points defined for any Desarguesian affine space $(\mathscr{P}, \mathscr{L}, \mathscr{E})$ is associative. In other words, (P + Q) + S = P + (Q + S) whenever P, Q, and S are in $\mathscr{P}$.

Proof: We will prove the typical case where no three of the points O, P, Q, and S are collinear (see Fig. 6.9). The following pairs of lines are parallel:

$\ell(O, P) \| \ell(Q, P + Q)$ $\ell(P, P + Q) \| \ell(O, Q)$

$\ell(O, S) \| \ell(Q, Q + S)$ $\ell(O, P + Q) \| \ell(S, (P + Q) + S)$

$\ell(O, S) \| \ell(P + Q, (P + Q) + S)$

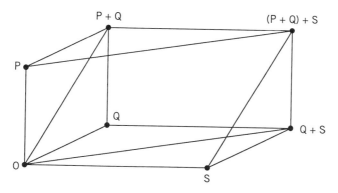

Figure 6.9

Thus $\ell(Q, Q + S) \| \ell(P + Q, (P + Q) + S)$ by $AS4$. Applying Desargues' Theorem to $O, Q, P + Q$, and $S, Q + S, (P + Q) + S$ gives $\ell(Q, P + Q) \| (\ell(Q + S, (P + Q) + S)$. By the transitivity of parallelism it follows that $\ell(O, P) \| \ell(Q + S, (P + Q) + S)$. Desargues' Theorem, this time applied to $P, P + Q, (P + Q) + S$, and $O, Q, Q + S$, gives $\ell(P, (P + Q) + S) \| \ell(O, Q + S)$. Thus $(P + Q) + S$ satisfies the condition in the definition of the sum $P + (Q + S)$, and the two are equal.

The proofs of the other (exceptional) cases are left as exercises. ■

The next two theorems complete the proof that $\mathscr{P}$ is an abelian group under addition.

■ **THEOREM 8.** $O + P = P$ for any P in $\mathscr{P}$. For each P in $\mathscr{P}$ there is a point $(-P)$ such that $P + (-P) = (-P) + P = O$.

Proof: Since P and O are collinear, the result follows from Theorems 2, 3, and 7 of Chapter 3. ■

■ **THEOREM 9.** For any points P and Q in $\mathscr{P}$, $P + Q = Q + P$.

Proof: If P, Q, and O are all on some line m, then by Theorem 11 of Chapter 3, addition is commutative on (m, O, I_m), and the result follows. If P, Q, and O are noncollinear, then $(O, P, P + Q, Q)$ is a parallelogram if and only if $(O, Q, P + Q, P)$ is a parallelogram, so addition is commutative. ■

Definition. Let $a \in \mathbb{D}$ and $P \in \mathscr{P}$. Then aP is defined to be the point in $\ell(O, P)$ which is the product (as defined in Chapter 3) of $\phi_{\ell(O, P)}(a)$ and P in the ring $(\ell(O, P), O, \phi_{\ell(O, P)}(1))$, as shown in Fig. 6.10.

Desargues' Theorem can be used to show that scalar multiplication is independent of the choice of B (see Exercise 3b below). It follows

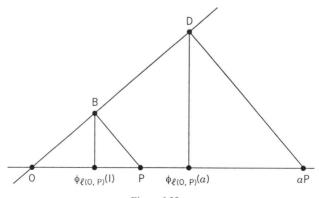

Figure 6.10

immediately from the definition of scalar multiplication that 1P is the product of $\phi_{\ell(O, P)}(1)$ and P, which is $I_{\ell(O, P)} \cdot P = P$.

The associative and distributive laws for scalar multiplication can now be derived to complete the proof that $\mathscr{P}$ is a vector space over $\mathbb{D}$.

■ **THEOREM 10.** For a and b in $\mathbb{D}$, with P and Q in $\mathscr{P}$,

   i. $a(P + Q) = aP + aQ$,
   ii. $(a + b)P = aP + bP$,
   iii. $(ab)P = a(bP)$.

Proof: If O, P, and Q are collinear, statement i is true by distributivity of multiplication over addition in the ring $\ell(O, P)$.

Otherwise, let $\ell_1 = \ell(O, P)$, $\ell_2 = \ell(O, Q)$, and $\ell_3 = \ell(O, P + Q)$, as shown in Fig. 6.11. Let $\phi_{\ell_i}(a) = A_i$, and $\phi_{\ell_i}(1) = I_i$ for $i = 1, 2, 3$. Then $A_1 P = aP$, $A_2 Q = aQ$, and $A_3 (P + Q) = a(P + Q)$.

Multiply aP using I_3 (and A_3), multiply aQ using I_3 (and A_3), and multiply $A_3(P + Q)$ using P (and aP). Thus $\ell(P, P + Q) \| \ell(aP, a(P + Q))$, whence $\ell(O, Q) \| \ell(aP, a(P + Q))$ by the transitivity of parallelism. Desargues' Theorem on P, I_3, Q, and aP, A_3, aQ, yields $\ell(P, Q) \| \ell(aP, aQ)$. Desargues' Theorem on $P, Q, P + Q$ and $aP, aQ, a(P + Q)$ gives $\ell(Q, P + Q) \| \ell(aQ, a(P + Q))$. Thus $\ell(O, P) \| \ell(aQ, a(P + Q))$. This statement, together with the result that $\ell(O, Q) \| \ell(aP, a(P + Q))$ implies that

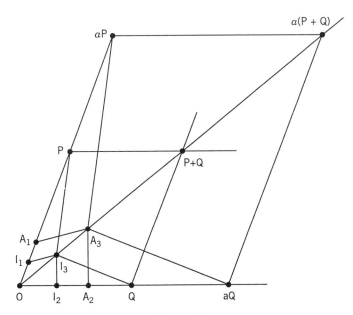

Figure 6.11

aP + aQ = a(P + Q). Thus, statement i holds when O, P, and Q are noncollinear.

Since addition and multiplication here take place on ℓ(O, P), statement ii follows from the distributive laws proved in Theorems 18 and 19 of Chapter 3. Statement iii follows from the associative law proved in Theorem 16 of Chapter 3.   ■

The results above can be summarized in the following theorem:

■   **THEOREM 11.** Relative to O and the points I_ℓ selected as identities above, $\mathscr{P}$ is a vector space over $\mathbb{D}$.

Having shown that any affine space can be considered as a vector space, it can now be proved that the lines in $\mathscr{L}$ correspond to the lines of the vector space as defined in Example 2 of Section 6.1 above.

■   **THEOREM 12.** For P and Q distinct points of the Desarguesian affine space $(\mathscr{P}, \mathscr{L}, \mathscr{E})$,

$$\ell(\text{P}, \text{Q}) = \{\text{S} = \text{P} + t(\text{Q} - \text{P}) : t \in \mathbb{D}\}$$

Proof:

CASE 1. Suppose that O, P, and Q are collinear. Then, if S = P + t(Q − P), S ∈ ℓ(O, P) = ℓ(P, Q), since all addition and multiplications here take place on that line. Conversely, if S ∈ ℓ(P, Q), then take T = (S − P)(Q − P)$^{-1}$ in ℓ(P, Q). Since T = $\phi_{\ell(\text{O}, \text{P})}(t)$ for some t in $\mathbb{D}$, it follows that S − P = T(Q − P) = t(Q − P), and S = P + t(Q − P).

CASE 2. Suppose that O is not in ℓ(P, Q), as shown in Fig. 6.12, and let S ∈ ℓ(P, Q). Now ℓ(O, P)∥ℓ(Q − P, Q) and ℓ(P, Q)∥ℓ(O, Q − P). Then there exists a point X with ℓ(O, X)∥ℓ(P, S) and ℓ(O, P)∥ℓ(X, S). Thus, X + P = S. But X is in ℓ(P, Q − P) by the parallel axiom; hence X = t(Q − P) for some t in $\mathbb{D}$. Thus S = P + t(Q − P).

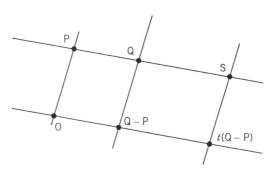

Figure 6.12

Conversely, if $S = P + t(Q - P)$, we have $\ell(P, S)\|\ell(O, t(Q - P))$. But $\ell(O, Q - P)\|\ell(P, Q)$ and $\ell(O, Q - P) = \ell(O, t(Q - P))$, so $\ell(P, S)\|\ell(O, Q - P)$. By the parallel axiom $\ell(P, S) = \ell(P, Q)$, so $S \in \ell(P, Q)$. ∎

Corollary. For P and Q distinct points of $\mathscr{P}$, $\ell(P, Q) = \{tP + (1 - t)Q : t \in \mathbb{D}\}$. For $P \neq O$, $\ell(O, P) = \{tP : t \in \mathbb{D}\}$.

The proof follows immediately from Theorem 12.

■ **THEOREM 13.** For P, Q, and S three noncollinear points of a Desarguesian affine space $(\mathscr{P}, \mathscr{L}, \mathscr{E})$, $\{\overline{PQS}\} = \{X = P + t(Q - P) + t'(S - P) : t, t' \in \mathbb{D}\}$.

Proof: Any point in the set on the right (which we will call $\mathscr{T}$) is clearly in $\{\overline{PQS}\}$, since P, S, and $P + t(Q - P)$ are in that plane, and $\ell(P, S)\|\ell(P + t(Q - P), P + t(Q - P) + t'(S - P))$, since both are parallel to $\ell(O, S - P)$, as shown in Fig. 6.13.

To obtain the reverse inclusion, it is enough to show that $\mathscr{T}$ is a flat of $(\mathscr{P}, \mathscr{L}, \mathscr{E})$, since P, Q, and S are contained in it. Let $X = P + t(Q - P) + t'(S - P)$, $Y = P + r(Q - P) + r'(S - P)$, and $Z \in \ell(X, Y)$. Then $Z = X + s(Y - X) = P + t(Q - P) + t'(S - P) + s((r - t)(Q - P) + (r' - t')(S - P)) + (t + s(r - t))(Q - P) + (t' + s(r' - t'))(S - P)$, which is clearly in $\mathscr{T}$. ∎

Corollary. For P, Q, and S three noncollinear points of $\mathscr{P}$, $\{\overline{PQS}\} = \{tP + rQ + (1 - t - r)S : t, r \in \mathbb{D}\}$. If O, P, and Q are noncollinear, then $\{\overline{OPQ}\} = \{tP + rQ : t, r \in \mathbb{D}\}$.

The corollary follows immediately from Theorem 13. The connection between affine flats and linear subspaces will be established next.

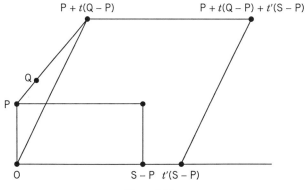

Figure 6.13

■   **THEOREM 14.** Let $\mathscr{F}$ be an affine flat of the Desarguesian affine space $(\mathscr{P}, \mathscr{L}, \mathscr{E})$ such that $O \notin \mathscr{F}$. Then for each point P in $\mathscr{F}$, $\mathscr{S} = \{Q - P : Q \in \mathscr{F}\}$ is an affine flat containing O. Furthermore $\mathscr{S}$ is a linear subspace of the vector space $\mathscr{P}$ and $\mathscr{F} = \{S + P : S \in \mathscr{S}\}$.

Proof: For Q and T in $\mathscr{F}$ with r with $\mathbb{D}$, if $X \in \ell(Q - P, T - P)$, then $X = (Q - P) + r((T - P) - (Q - P)) = Q - P + r(T - Q) = [Q + r(T - Q)] - P$. But $Q + r(T - Q) \in \mathscr{F}$, so $X \in \mathscr{S}$ and $\mathscr{S}$ is a subspace. $\mathscr{S}$ clearly contains $P - P = O$, and $\mathscr{F} = \{S + P : S \in \mathscr{S}\}$.

To show that $\mathscr{S}$ is a linear subspace of $\mathscr{P}$, consider $Q - P$ and $T - P$ in $\mathscr{S}$ (where P, Q, and T are in $\mathscr{F}$). Then $(Q - P) + (T - P) = (Q + T - P) - P$. But $Q + T - P$ is the fourth vertex of the parallelogram with vertices Q, T, and P, and so is in the affine flat $\mathscr{F}$. Thus $(Q - P) + (T - P)$ is in $\mathscr{S}$. For $Q - P$ in $\mathscr{S}$ and any t in $\mathbb{D}$, $t(Q - P) = (P + t(Q - P)) - P$. since $P + t(Q - P)$ is in $\mathscr{F}$, $t(Q - P)$ is in $\mathscr{S}$ and $\mathscr{S}$ is a vector subspace of $\mathscr{P}$.   ■

In general, if $\mathscr{S}$ is a linear subspace of a vector space $\mathscr{V}$ and $\mathscr{F} = \{x + v : x \in \mathscr{S}\}$ for some fixed v in $\mathscr{V}$, $\mathscr{F}$ is called a *translate* of $\mathscr{S}$. Thus the affine flats of any Desarguesian space $(\mathscr{P}, \mathscr{L}, \mathscr{E})$ correspond to the translates of the linear subspaces of the vector space $\mathscr{P}$. Such a translate is itself a vector subspace if and only if it contains the origin of the vector space.

EXERCISES

1. In a Desarguesian affine space with m and k lines containing O, prove that $\ell(\phi_m(a), \phi_k(a)) \| \ell(\phi_m(1), \phi_k(1))$ whenever a is an element of $\mathbb{D}$ different from 0 or 1.

2. Let $(\mathscr{P}, \mathscr{L}, \mathscr{E})$ be the eight-point affine space given in Example 3 above. Relative to $C = O$, add the points G and E.

3. a. Use Desargues' Theorem to show that addition of points P and Q, collinear with O in a Desarguesian affine space, is independent of the choice of the auxiliary point B.

   b. Do the same for scalar multiplication of a ring element a times a point P.

4. a. Suppose that O, P, and Q are collinear points of any Desarguesian affine space $(\mathscr{P}, \mathscr{L}, \mathscr{E})$, with S not on $\ell(O, P)$. Prove that $(P + Q) + S = P + (Q + S)$.

   b. Suppose that P, Q, and S are collinear points of any Desarguesian affine space $(\mathscr{P}, \mathscr{L}, \mathscr{E})$, with O not on $\ell(P, Q)$. Prove that $(P + Q) + S = P + (Q + S)$.

5. Let $(\mathscr{P}, \mathscr{L}, \mathscr{E})$ be a Desarguesian affine space with (A, B, C, D) a parallelogram (i.e., $\ell(A, B) \| \ell(C, D)$ and $\ell(A, D) \| \ell(B, C)$). Suppose that O is not on any of the sides of the parallelogram. Prove that $A + C = B + D$.

6. Show that $(\mathscr{P}, \mathscr{L}, \mathscr{E})$ satisfies Pappus' Theorem, as discussed in Section 3.7, if and only if its coordinatizing ring $\mathbb{D}$ is a field.

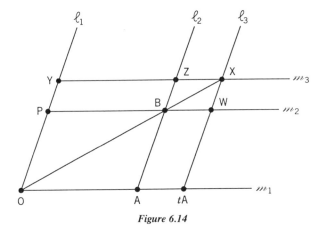

Figure 6.14

7. Let $(\mathscr{P}, \mathscr{L}, \mathscr{E})$ be an affine space, and let $\ell_1 \| \ell_2 \| \ell_3$ and $m_1 \| m_2 \| m_3$ with the lines intersecting in the nine points as shown in Fig. 6.14. Suppose that B, O, and X are collinear. All points shown are in the plane $\{\overline{OAB}\}$. Thus each point may be written as $rA + sB$ for some ring elements r and s. Write Y, Z, X, P, and W in this form using A and B from $\mathscr{P}$, and $0, \pm 1, \pm t$, from $\mathbb{D}$. (Do *not* assume that the space contains only nine points.)

SUGGESTED READING

U. Sasaki, Lattice theoretic characterization of an affine geometry of arbitrary dimensions. *J. Sci. Hiroshima U.* **16** (1952), 226.

CHAPTER 7

Projective Space

The connection between affine and projective planes extends to affine spaces. As in the planar case, a "projective space" can be obtained by augmenting an affine space. This chapter describes that construction, and the coordinatization of projective spaces.

7.1 SYNTHETIC PROJECTIVE SPACE

In this section we present some examples of projective space together with examples showing that the axioms are independent.

Definition. A *projective space* is an ordered pair $(\mathscr{P}, \mathscr{L})$, where $\mathscr{P}$ is a nonempty set whose elements are called *points* and $\mathscr{L}$ is a nonempty collection of subsets of $\mathscr{P}$ called *lines* subject to the following axioms:

*PS*1. Given any two distinct points P and Q, there is one and only one line (called $\ell(P, Q)$) containing them.

*PS*2. (The Pasch Axiom) If A, B, C, and D are distinct points such that there is a point E in $\ell(A, B) \cap \ell(C, D)$, then there is a point F in $\ell(A, C) \cap \ell(B, D)$ (see Fig. 7.1).

*PS*3. Each line contains at least three points; not all points are collinear.

The only difference between the axioms for a projective plane and those for a projective space is the second axiom, due to Moritz Pasch (1843–1930) (but in a different context) and first used in projective space by the American mathematician Oswald Veblen (1880–1960) as a clever way to avoid postulating the existence of planes as well as points and lines. This axiom postulates the intersection of *certain* lines in a projective

144

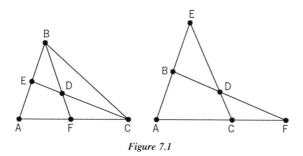

Figure 7.1

space, while *any* two lines in a projective plane intersect. Intuitively, if $\ell(A, B)$ and $\ell(C, D)$ intersect, then they would seem to be in some plane, so $\ell(A, C)$ and $\ell(B, D)$ would be coplanar and therefore intersect as well. In the next section planes in projective spaces will be defined, and it will be shown that two lines intersect if and only if they are coplanar.

After giving some examples and results about projective spaces, including the projective version of Desargues' Theorem, the construction of an affine space from a projective space will be described, as well as the reverse construction. As in the planar case the coordinatization of affine spaces will be used to coordinatize projective spaces. Also as in the planar case it is easier to make *geometric proofs* in *projective* spaces but easier to *coordinatize affine* spaces.

Some examples will show the consistency and independence of the axioms for projective space.

□**EXAMPLE 1:** If $(\mathscr{P}, \mathscr{L})$ is any projective plane (e.g., the seven-point, seven-line Fano plane given in Example 5 of Chapter 2, and discussed in Example 6 of Chapter 6), then $(\mathscr{P}, \mathscr{L})$ is a projective space. □

□**EXAMPLE 2: The Fano space.** The Fano plane is one of the planes in this example, discovered by Gino Fano in 1892. Robinson (1990) describes the Fano space as consisting of 15 points—represented by the symbols $(ab), (ac), (ad), (ae), (af), (bc), (bd), (be), (bf), (cd), (ce), (cf), (de), (df),$ (ef), with $(ij) = (ji)$— and 35 lines: 15 of the form $\{(ab), (bc), (ca)\}$ and 20 of the form $\{(ab), (cd), (ef)\}$. In the Fano space there are 15 planes, each of which has seven points and seven lines. The Fano space was the first example given of a finite projective space. □

Examples 1 and 2 show that the axioms for projective space are *consistent*.

□**EXAMPLE 3: Coordinate projective spaces.** A whole class of examples of projective spaces can be given in terms of the vector spaces F^n (or $\mathbb{D}^n$), where F is any field (or $\mathbb{D}$ any division ring). As was done for F^3 in Section 5.1, define an equivalence relation $\sim$ on the nonzero vectors of F^n by $(x_1, \ldots, x_n) \sim (y_1, \ldots, y_n)$ if and only if $(x_1, \ldots, x_n) = (ty_1, \ldots, ty_n)$ for some nonzero $t \in F$, and let $\langle \mathbf{x} \rangle$ or $\langle x_1, \ldots, x_n \rangle$ represent the equivalence class of $\mathbf{x}$, where $\mathbf{x} = (x_1, \ldots, x_n)$. Then $(\mathscr{P}, \mathscr{L})$ is a projective space where $\mathscr{P} = \{\langle \mathbf{x} \rangle : \mathbf{x} \in F^n\}$, and the line containing $\langle \mathbf{x} \rangle$ and $\langle \mathbf{y} \rangle$ is the set of points of the form $\langle s\mathbf{x} + t\mathbf{y} \rangle$, where s and t take on all values in F. The projective space described here is called $P_{n-1}(F)$ or $PG(F^n)$, and projective spaces can be formed similarly when F is replaced by a division ring $\mathbb{D}$. These spaces, called *projective spaces over F (or $\mathbb{D}$), will be discussed in more detail below.*   □

   Further examples show that the axioms for projective space are independent.

□**EXAMPLE 4: The independence of PS1.** Let $\mathscr{P} = \{ABCD\}$ and $\mathscr{L} = \{\{ABC\}, \{ADC\}\}$. Here A and C are on *two* lines, so *PS*1 fails. *PS*2 holds, since any two lines intersect, and *PS*3 clearly holds.   □

□**EXAMPLE 5:** Euclidean three-space satisfies *PS*1 and *PS*3 but not *PS*2.   □

□**EXAMPLE 6:** If $\mathscr{P} = \{ABC\}$ and $\mathscr{L} = \{\{AB\}, \{AC\}, \{BC\}\}$, then the first two axioms hold but the third one fails.   □

EXERCISES

1. Prove that any two lines in a projective space have the same number of points.

2. Are the projective points $\langle 1, 2, 2, 4 \rangle$, $\langle 3, 1, 2, 5 \rangle$, and $\langle 4, 2, -1, 1 \rangle$ collinear in the real projective space $P_3(\mathbb{R})$?

3. If possible, find the point of intersection of the projective lines $\ell(\langle 1, 3, 5, 2 \rangle, \langle 3, 4, 2, 1 \rangle)$ and $\ell(\langle 1, 1, 2, 1 \rangle, \langle 1, 4, 3, 2 \rangle)$ in $P_3(\mathbb{R})$.

4. Are the points $\langle 1, 1, 1, 1 \rangle$, $\langle 0, u, u + 1, u \rangle$ and $\langle u, 1, 0, u + 1 \rangle$ collinear in $P_3(GF(4))$? Why or why not?

5. How many points are in the projective space $P_n(\mathbb{Z}_2)$? How many lines?

6. How many points and lines are in the projective space $P_3(\mathbb{Z}_p)$ where p is a prime number?

7. Prove that the projective points $\langle \mathbf{x} \rangle$ in $PG(F^n)$ are in 1-1 correspondence with the affine lines through the origin in the affine space $AG(F^n)$.

*8. Prove that the projective lines in $PG(F^n)$ are in 1-1 correspondence with the affine planes through the origin in $AG(F^n)$.

7.2 PLANES IN PROJECTIVE SPACE

The next goal is to define the concept of a *plane* in a projective space, and to show that such planes are the same as the *projective planes* discussed in Section 2.6. In particular, any two lines in a plane intersect, and conversely, intersecting lines must be coplanar. It will be shown that planes are *flats*, meaning that together with any two points, a plane contains their line, and that any three noncollinear points are contained in exactly one plane. This discussion is just the first step in the development of the theory of general projective flats, and it will extend to include a notion of *dimension* in projective space.

Definition. Let M be a point of $(\mathscr{P}, \mathscr{L})$ not in line ℓ. The *plane* α determined by M and ℓ is given by

$$\alpha = \bigcup_{A \in \ell} \ell(M, A) = \{P : \ell(P, M) \cap \ell \neq \varnothing\} \cup \{M\}.$$

We will denote this plane by (M, ℓ).

■ **THEOREM 1.** If T and P are points in plane (M, ℓ), then $\ell(T, P) \subseteq (M, \ell)$.

Proof: Suppose that $R \in \ell(T, P)$, and let $\ell(T, M) \cap \ell = Q$ while $\ell(P, M) \cap \ell = S$. It must be shown that $\ell(R, M) \cap \ell(Q, S) \neq \varnothing$. First, by the Pasch Axiom, since $\ell(Q, T) \cap \ell(S, P) = M$, then $\ell(Q, S) \cap \ell(T, P) = X$ for some point X (see Fig. 7.2). Again by the Pasch Axiom, since $\ell(X, R) \cap \ell(M, S) = P$, then $\ell(R, M) \cap \ell(X, S) \neq \varnothing$. But $\ell(X, S) = \ell$; hence $R \in (M, \ell)$. (We will here and in later proofs omit trivial cases such as $R = M$.)

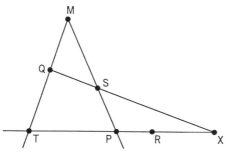

Figure 7.2

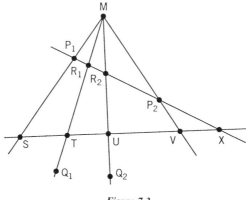

Figure 7.3

■ **THEOREM 2.** Two lines in a projective space intersect if and only if they are coplanar.

Proof: (→): If $\ell(P, Q) \cap \ell(R, S) = T$, then both lines are on the plane $(T, \ell(P, R))$.

(←): Suppose that P_1, P_2, Q_1, and Q_2 are in plane (M, ℓ), and let S, T, U, and V be the intersections of ℓ with $\ell(M, P_1)$, $\ell(M, Q_1)$, $\ell(M, Q_2)$, and $\ell(M, P_2)$, respectively, as shown in Fig. 7.3. Now use the Pasch Axiom four times:

1. $\ell(S, P_1) \cap \ell(V, P_2) = M$; thus $\ell(P_1, P_2) \cap \ell(S, V) = X$.
2. $\ell(P_1, M) \cap \ell(X, T) = S$; thus $\ell(P_1, X) \cap \ell(M, T) = R_1$.
3. $\ell(P_2, M) \cap \ell(X, U) = V$; thus $\ell(P_2, X) \cap \ell(M, U) = R_2$.
4. $\ell(Q_1, R_1) \cap \ell(Q_2, R_2) = M$; thus $\ell(Q_1, Q_2) \cap \ell(R_1, R_2) \neq \varnothing$.
   Since $\ell(R_1, R_2) = \ell(P_1, P_2)$, the proof is finished.   ■

Corollary. Any plane in a projective space is a projective plane.

■ **THEOREM 3.** Suppose that $K \in (M, \ell)$ with K not in ℓ. Then $(K, \ell) = (M, \ell)$.

Proof: If $P \in (M, \ell)$, then $\ell(K, P) \cap \ell = X$ for some X by Theorem 2. Thus $\ell(K, X)$ contains P, so P is in (K, ℓ) and $(M, \ell) \subseteq (K, \ell)$. A symmetric argument shows that $(K, \ell) \subseteq (M, \ell)$.   ■

■ **THEOREM 4.** Any three noncollinear points in a projective space lie in a unique plane.

Proof: Suppose that M, N, and K are noncollinear in $(\mathscr{P}, \mathscr{L})$. Take P in $\ell(M, N)$, P different from M and N, and take Q in $\ell(M, K)$, Q different from M and K. Then $(M, \ell(N, K)) = (M, \ell(P, Q)) = (K, \ell(P, Q)) = (K, \ell(M, N))$. Similarly $(M, \ell(N, K)) = (N, \ell(M, K))$. If M, N, and K are in some plane $(X, \ell(Y, Z))$, assume without loss of generality that $X \notin \ell(N, K)$. Then $(X, \ell(Y, Z)) = (X, \ell(N, K)) = (M, \ell(N, K))$, and the proof is complete. ■

Corollary. Given a point M not contained in a line ℓ, then (M, ℓ) is the unique plane containing M and ℓ.

It was shown in Theorem 1 that planes in projective space satisfy the defining condition for flats as described for affine space in Section 6.2.

Definition. Let $(\mathscr{P}, \mathscr{L})$ be a projective space with $\mathscr{M}$ a subset of $\mathscr{P}$. $\mathscr{M}$ is a *flat* of $(\mathscr{P}, \mathscr{L})$ if, whenever P and Q are distinct points in $\mathscr{M}$, $\ell(P, Q) \subseteq \mathscr{M}$. If $\mathscr{N}$ is a subset of $\mathscr{P}$, its *projective closure* $\overline{\mathscr{N}}$ is the intersection of all flats containing $\mathscr{N}$.

The mapping taking any subset of $\mathscr{P}$ to its projective closure is a closure operator on $\mathscr{P}$ (see Exercise 4d).

■ **THEOREM 5.** Let $\mathscr{M}$ be a flat of the projective space $(\mathscr{P}, \mathscr{L})$, and let Q be a point not in $\mathscr{M}$. Then

$$\overline{\{Q\} \cup \mathscr{M}} = \bigcup_{A \in \mathscr{M}} \ell(Q, A).$$

Proof: (see Fig. 7.4). Clearly $\{Q\} \cup \mathscr{M} \subseteq \bigcup_{A \in \mathscr{M}} \ell(Q, A) \subseteq \overline{\{Q\} \cup \mathscr{M}}$ so if the middle set (which we will call $\mathscr{U}$) can be shown to be a flat, it must be equal to $\overline{\{Q\} \cup \mathscr{M}}$. Thus we take $D \in \ell(B, C)$, where $B \in \ell(Q, A)$ and $C \in \ell(Q, E)$; A and E are points in $\mathscr{M}$. Then Q, D, A, and E are coplanar (in $(Q, \ell(B, C))$, whence $\ell(A, E) \cap \ell(Q, D) = F \in \mathscr{M}$. Since $D \in \ell(Q, F)$, $D \in \mathscr{U}$, and $\mathscr{U}$ is a flat. ■

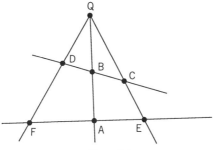

Figure 7.4

EXERCISES

1. Show that, in any projective space, the intersection of a line and a plane not containing it is either empty or a point.

2. In the Fano space given in Example 2, list the points on the plane determined by the point (cf) and the line $\{(ab), (cd), (ef)\}$.

3. Suppose that $k \subseteq (M, \ell)$ with $M \notin k$. Prove that $(M, \ell) = (M, k)$.

4. Let $(\mathscr{P}, \mathscr{L})$ be a projective space. Prove each of the following statements:

   a. Each line in $\mathscr{L}$ is a flat.

   b. Each plane in $(\mathscr{P}, \mathscr{L})$ is a flat.

   c. The intersection of an arbitrary (finite or infinite) collection of flats is again a flat.

   d. The mapping taking each subset of $\mathscr{P}$ to its projective closure is a closure operator (as defined in Chapter 6) on $\mathscr{P}$.

5. Suppose $\mathscr{M}$ and $\mathscr{N}$ are flats of the projective space $(\mathscr{P}, \mathscr{L})$. Show that $\overline{\mathscr{M} \cup \mathscr{N}} = \bigcup_{\substack{A \in \mathscr{M} \\ B \in \mathscr{N}}} \ell(A, B)$.

7.3 DIMENSION

It follows from Theorem 5 that the plane (M, ℓ) is simply the flat $\overline{\{M\} \cup \ell}$ and that any flat and point not containing it together determine a larger flat—intuitively, a flat one dimension higher than the original one. The concept of dimension in projective space can be defined *recursively* as follows:

   Definition. A point in projective space is called a 0-*dimensional flat* (or 0-*flat*) and a line is a 1-*flat*. In general, when $\mathscr{M}$ is an n-flat in projective space and Q is a point not in $\mathscr{M}$, the flat $\overline{\{Q\} \cup \mathscr{M}}$ is an $(n + 1)$-*flat*.

   As expected, this defines planes as 2-flats, while a 3-flat (or 3-*space*) is determined by a plane together with a point not on it. To complete the definition, the empty set is called a (-1)-dimensional flat. For convenience, we will continue to refer to the plane determined by M and ℓ as (M, ℓ), and we will denote the 3-space given by point M and plane α by (M, α).

■ **THEOREM 6.** The *exchange theorem*. Let $\mathscr{M}$ be a flat of $(\mathscr{P}, \mathscr{L})$ not containing points P and Q. If $P \in \overline{\{Q\} \cup \mathscr{M}}$, then $\overline{\{P\} \cup \mathscr{M}} = \overline{\{Q\} \cup \mathscr{M}}$.

   Proof: $P \in \ell(Q, A)$ for some A in $\mathscr{M}$. Thus $Q \in \ell(P, A)$, whence $Q \in \overline{\{P\} \cup \mathscr{M}}$, and $\overline{\{Q\} \cup \mathscr{M}} \subseteq \overline{\{P\} \cup \mathscr{M}}$. The reverse inclusion is trivial, so equality obtains.   ■

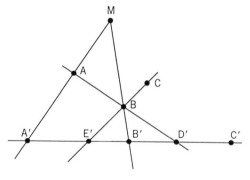

Figure 7.5

■ **THEOREM 7.** Suppose that β is a plane contained in the three-space (M, α), such that $M \notin \beta$. Then $(M, \alpha) = (M, \beta)$.

Proof: Since $\{M\} \cup \beta \subseteq (M, \alpha) = \overline{\{M\} \cup \alpha}$, it follows that $\overline{\{M\} \cup \beta} \subseteq \overline{\{M\} \cup \alpha}$. To show the reverse inclusion, let $\beta = \overline{\{A, B, C\}}$, and let $A' = \angle(M, A) \cap \alpha, B' = \angle(M, B) \cap \alpha$, and $C' = \angle(M, C) \cap \alpha$. If A', B', and C' are noncollinear, then the plane they determine is α. Since $A \in \angle(M, A') \cap \beta$, $B \in \angle(M, B') \cap \beta$, and $C \in \angle(M, C') \cap \beta$, it follows that $\alpha \subseteq (M, \beta)$. Therefore $\overline{\{M\} \cup \alpha} \subseteq \overline{\{M\} \cup \beta}$.

If A', B', and C' are collinear, as shown in Fig. 7.5, then since $\angle(A', A) \cap \angle(B', B) = M$, $\angle(A, B) \cap \angle(A', B') = D'$ for some D' and similarly $\angle(B, C) \cap \angle(B', C') = E'$ for some E'. But this implies that both D' and E' are in β so that $\angle(D', E') \subseteq \beta$, and in particular $A' \in \beta$. Thus $M \in \angle(A, A') \subseteq \beta$, a contradiction. Therefore A', B', and C' are noncollinear, and the result follows. ■

■ **THEOREM 8.** Let α and β be distinct planes with β contained in (M, α). Then α and β intersect in a line.

Proof: Suppose that A, B, and C are three noncollinear points of β (see Fig. 7.6). Then $\angle(M, A) \cap \alpha = D$, $\angle(M, B) \cap \alpha = E$ and $\angle(M, C) \cap \alpha = F$. Since A, B, D, and E are coplanar [in $(M, \angle(A, B)]$, $\angle(A, B) \cap \angle(D, E) = G \in \alpha \cap \beta$; similarly $\angle(B, C) \cap \angle(E, F) = H \in \alpha \cap \beta$. Consequently $\angle(G, H) \subseteq \alpha \cap \beta$, and by Theorem 4 there cannot be a point on both α and β that is not on $\angle(G, H)$. Therefore $\angle(G, H) = \alpha \cap \beta$. ■

■ **THEOREM 9.** If two distinct planes are contained in a 3-space, they intersect in a line.

Proof: Let α and β be subsets of the 3-space (X, ∂). Assuming that neither α nor β equals ∂, then $\alpha \cap \partial = \ell$ and $\beta \cap \partial = m$ for some lines ℓ and m. Since ℓ and m are in plane ∂, they intersect in a point T. If

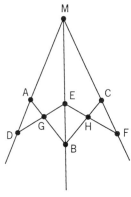

Figure 7.6

$X \in \alpha \cap \beta$, then $\ell(T, X)$ is in $\alpha \cap \beta$. Otherwise, we can assume that $X \notin \beta$, so $(X, \partial) = (X, \beta)$. Thus $\alpha \subseteq (X, \beta)$, and by Theorem 8, α and β intersect in a line.   ∎

We can now prove Desargues' Theorem in any nonplanar projective space. The statement of the theorem is the same as that given for the projective plane in Chapter 5. As in the proof that Desargues' Theorem holds in every nonplanar affine space (Theorem 4 of Chapter 6), there are two cases, and the first is used to prove the second. Note, however, that Desargues' Theorem is a single statement in the projective case, whereas in the affine case it has two parts, depending on whether the lines through the corresponding vertices of the triangles are parallel or intersect.

■ **THEOREM 10. Desargues' Theorem for projective space.** Let $(\mathscr{P}, \mathscr{L})$ be a projective space in which not all points are coplanar. Let A, B, C and A′, B′, C′ be two triples of noncollinear points such that $\ell(A, A') \cap \ell(B, B') \cap \ell(C, C') = P$. Suppose that $\ell(A, B) \cap \ell(A', B') = C''$, $\ell(A, C) \cap \ell(A', C') = B''$, and $\ell(B, C) \cap \ell(B', C') = A''$. Then A″, B″, and C″ are collinear.

Proof: First note that A, B, A′, and B′ are in the plane $(P, \ell(A, B))$, so C″ exists. Similarly A″ and B″ exist.

CASE 1. $\alpha = (A, \ell(B, C)) \neq (A', \ell(B', C')) = \beta$. These planes are both in the three space (P, α); thus they intersect in a line. But A″, B″, and C″ are in both α and β; therefore they are collinear (see Fig. 7.7).

CASE 2. $(A, \ell(B, C)) = (A', \ell(B', C')) = \alpha$. Select Q not in α and R in $\ell(P, Q)$. (R is not in α since P is in α.) Let $A^\sim = \ell(Q, A') \cap \ell(R, A)$, $B^\sim = \ell(Q, B') \cap \ell(R, B)$, and $C^\sim = \ell(Q, C') \cap \ell(R, C)$, as shown in Fig. 7.8. Now Q, A′, R, and A are on the plane $(P, \ell(R, A'))$, so $A^\sim$ exists; similarly $B^\sim$ and $C^\sim$ exist. Let β be the plane $(A^\sim, \ell(B^\sim, C^\sim))$. Now use Case 1 on triangles A, B, C and $A^\sim$, $B^\sim$, $C^\sim$ to get D, E, and F

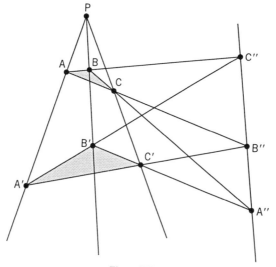

Figure 7.7

collinear on the line $\alpha \cap \beta = m$, where $\ell(A, C) \cap \ell(A^\sim, C^\sim) = D$, $\ell(A, B) \cap \ell(A^\sim, B^\sim) = E$, and $\ell(B, C) \cap \ell(B^\sim, C^\sim) = F$. Use Case 1 on triangles A', B', C' and $A^\sim$, $B^\sim$, $C^\sim$ to get G, H, and J collinear on m, where $\ell(A', C') \cap \ell(A^\sim, C^\sim) = G$, $\ell(A', B') \cap \ell(A^\sim, B^\sim) = H$, and $\ell(B, C) \cap \ell(B^\sim, C^\sim) = J$. Thus D and G are both in $m \cap \ell(A^\sim, B^\sim)$, so $D = G = \ell(A, C) \cap \ell(A', C') = B''$. Similarly $E = H = \ell(A, B) \cap$

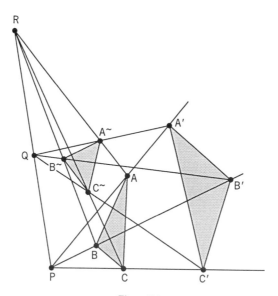

Figure 7.8

$\angle(A', B') = C''$, and $F = J = \angle(B, C) \cap \angle(B', C') = A''$. Therefore A'', B'', and C'' are collinear. ∎

EXERCISES

1. In $P_3(\mathbb{R})$, let $P = \langle 0, 0, 0, 1 \rangle$, $A = \langle 1, 0, 0, 1 \rangle$, $A' = \langle 3, 0, 0, 1 \rangle$, $B = \langle 0, 1, 0, 1 \rangle$, $B' = \langle 0, 4, 0, 1 \rangle$, $C = \langle 0, 0, 1, 1 \rangle$, and $C' = \langle 0, 0, 5, 1 \rangle$. Show that these points satisfy the hypotheses for Desargues' Theorem, and find the points A'', B'', and C''. Show that these points are collinear.

2. Show that any point in $PG(F^4)$ can be represented uniquely as $\langle x_1, x_2, x_3, 1 \rangle$, $\langle x_1, x_2, 1, 0 \rangle$, $\langle x_1, 1, 0, 0 \rangle$, or $\langle 1, 0, 0, 0 \rangle$.

3. Extend the result in Exercise 2 to $PG(F^n)$.

4. a. Use the results in Exercises 2 and 3 to count the number of points in $PG(F^n)$, where $F = GF(p^k)$.

   b. Calculate the number of lines in $PG(F^n)$, where $F = GF(p^k)$.

*5. a. Let $\mathscr{V} = \{f : \mathbb{Z}^+ \to \mathbb{R} : f(n) = 0$ for all but a finite number of elements of $\mathbb{Z}^+\}$, where $\mathbb{Z}^+$ is the set of positive integers $\{1, 2, 3, \dots\}$. Show that $\mathscr{V}$ is a vector space over $\mathbb{R}$. Show that $PG(\mathscr{V})$ has flats that are not finite dimensional. [*Hint*: Think of the elements of $\mathscr{V}$ as ∞-tuples, each with only a finite number of nonzero components.]

   b. Extend the results in (a) to an arbitrary field F.

7.4 CONSEQUENCES OF DESARGUES' THEOREM

Desargues' remarkable theorem leads to several observations. The only non-Desarguesian projective spaces are planes. Every projective space in which there are at least two planes is Desarguesian, and it will be shown in this section and the next that these spaces can be coordinatized by division rings.

Because the proof of Desargues' Theorem depends on the assumption that not all points are coplanar, it follows that neither the Moulton plane nor any other non-Desarguesian projective plane can be embedded in a higher-dimensional projective space. This was also the case in affine geometry as was seen in Chapter 6. The following consequence of Desargues' Theorem is a first step toward coordinatization.

■ **THEOREM 11.** Let $(\mathscr{P}, \mathscr{L})$ be a projective space such that not all points are coplanar. Then each plane of $(\mathscr{P}, \mathscr{L})$ can be coordinatized by a division ring.

Proof: Each plane of such a space is Desarguesian. ∎

It is possible to associate to each affine space a projective space in two different ways. The following construction of a projective space from an affine space is different from the correspondence between affine and projective planes described in Chapter 2, Theorem 13.

■   **THEOREM 12.** Let $(\mathscr{P}, \mathscr{L}, \mathscr{E})$ be an affine space such that not all points are coplanar. Let O be any point of $\mathscr{P}$. Then $(\mathscr{P}', \mathscr{L}')$ is a projective space where

$$\mathscr{P}' = \{\ell : \ell \in \mathscr{L} \text{ and } O \in \ell\}$$

and

$$\mathscr{L}' = \{\alpha : \alpha \in \mathscr{E} \text{ and } O \in \alpha\}.$$

(Note that if $(\mathscr{P}, \mathscr{L}, \mathscr{E})$ is an affine three space, then $(\mathscr{P}', \mathscr{L}')$ is a projective *plane*, and in general the projective space has dimension one less than that of the original affine space.)

Proof: If $\ell(O, P)$ and $\ell(O, Q)$ are distinct lines in $\mathscr{L}$, then O, P, and Q are noncollinear. Thus they determine a unique affine plane (i.e., a *line* in $\mathscr{L}'$) $\overline{\{OPQ\}}$, and *PS*1 holds. Any two elements of $\mathscr{L}'$ are of the form $\overline{\{OAB\}}$ and $\overline{\{OCD\}}$. By Theorem 5 of Chapter 6, if these affine planes intersect in the (projective) point $\ell(O, E)$, then there is an F such that $\ell(O, F) \subseteq \overline{\{OAC\}} \cap \overline{\{OBD\}}$, and the Pasch Axiom holds. The third axiom states that there are at least three points on each line. Thus it must be shown that any affine plane $\overline{\{OPQ\}}$ contains a point R such that $\ell(O, P)$, $\ell(O, Q)$, and $\ell(O, R)$ are distinct in $\mathscr{L}$. But this is clear by taking R to be the point such that $\ell(O, P) \| \ell(Q, R)$ with $\ell(O, Q) \| \ell(P, R)$. Since $\mathscr{P}$ is a nonplanar affine space, not all points in $\mathscr{P}'$ are collinear.   ■

In the vector space $\mathbb{D}^n$, a *one-dimensional* linear subspace consists of a nonzero vector and all its multiples, that is, an *affine* line in $AG(\mathbb{D}^n)$. Hence a one-dimensional linear subspace of $\mathbb{D}^n$ is a line of the form $\ell(\mathbf{0}, \mathbf{x}) = \{t\mathbf{x} : t \in \mathbb{D}\}$. A *two-dimensional* subspace is a plane through the origin of the affine space $AG(\mathbb{D}^n)$. Any line of the form $\ell(\mathbf{0}, \mathbf{x})$ corresponds to the projective point $\langle\mathbf{x}\rangle$ in $P_{n-1}(\mathbb{D})$, and any plane through $\mathbf{0}$ in the affine space $\mathbb{D}^n$ corresponds to a projective line in $PG(\mathbb{D}^n)$. Thus the construction of $(\mathscr{P}', \mathscr{L}')$ in Theorem 12 generalizes the construction of $PG(\mathbb{D}^n)$ from $\mathbb{D}^n$, and the proof of Theorem 12 shows that $PG(\mathbb{D}^n)$ satisfies the axioms for a projective space.

It was shown in Theorem 14 of Chapter 2 that an affine plane can be constructed from a projective one by removing a single line, together with all the points on that line. Higher-dimensional flats can be removed from higher-dimensional projective spaces to construct affine spaces as well. This leads to the definition of a *maximal* proper flat in a projective space.

Definition. Let $(\mathscr{P}, \mathscr{L})$ be a projective space. A flat $\mathscr{M}$ is called a *maximal proper flat* of $(\mathscr{P}, \mathscr{L})$ if $\mathscr{M} \neq \mathscr{P}$ and the only flat which properly contains $\mathscr{M}$ is $\mathscr{P}$.

It is clear that maximal proper *affine* flats can be defined in the same way as projective ones, namely as affine flats that are properly contained only in the flat consisting of all points. Every projective space contains maximal proper flats (in fact each point in any projective space is contained in a maximal proper flat). In reducing a projective plane to an affine one, a line (i.e., a maximal proper flat) was removed from the projective plane. Deleting a maximal proper flat from a general projective space gives an affine space, as the following sequence of results shows:

■ **THEOREM 13.** Let $\mathscr{M}$ be a maximal proper flat of $(\mathscr{P}, \mathscr{L})$, and let A and B be points not in $\mathscr{M}$. Then $\ell(A, B)$ intersects $\mathscr{M}$ in exactly one point.

Proof: Since $B \in \mathscr{P} = \overline{\mathscr{M} \cup \{A\}}$, by Theorem 5, $B \in \ell(A, M)$ for some $M \in \mathscr{M}$. Thus $M \in \mathscr{M} \cap \ell(A, B)$. If any other point were in this intersection, the whole line $\ell(A, B)$ would be a subset of $\mathscr{M}$, a contradiction. ■

■ **THEOREM 14.** Let $(\mathscr{P}', \mathscr{L}')$ be a projective space with $\mathscr{M}$ a maximal proper flat. Then $(\mathscr{P}, \mathscr{L}, \mathscr{E})$ is an affine space, where

$$\mathscr{P} = \{P : P \in \mathscr{P}', P \notin \mathscr{M}\} = \mathscr{P}' \setminus \mathscr{M}.$$

By definition, $\ell \in \mathscr{L}$ if $\ell = \ell' \cap \mathscr{P}$ for some ℓ' in $\mathscr{L}'$ that is not a subset of $\mathscr{M}$, and likewise $\alpha \in \mathscr{E}$ if $\alpha = \alpha' \cap \mathscr{P}$, where α' is a plane in $(\mathscr{P}', \mathscr{L}')$ that is not a subset of $\mathscr{M}$.

Proof: $AS1$ for affine space is immediate. To prove that the parallel axiom holds, take P in $\mathscr{P}$ but not in ℓ. By definition, there is an ℓ' such that $\ell' \cap \mathscr{M} = X$ and $\ell = \ell' \setminus \{X\}$. Now, if $m' = \ell'(P, X)$, then $m = m' \setminus \{X\} \in \mathscr{L}$, and $\ell \| m$, since the only point they had in common as projective lines was X, which has been removed from both lines. If any line k contains P and is parallel to ℓ, then, as projective lines, k' and ℓ' intersected at a point of $\mathscr{M}$. But $\ell' \cap \mathscr{M} = X$, whence k' contained P and X, and is therefore equal to $\ell'(P, X) = m'$. Thus m is the *only* line containing P and parallel to ℓ.

Axiom $AS3$ follows from the fact that any three noncollinear points in a projective space determine a unique plane in that space. To show the transitivity of parallelism, note that if $\ell \| m$ and $m \| k$, then $\ell' \cap \mathscr{M} = X = m' \cap \mathscr{M} = \ell' \cap m'$, and similarly $m' \cap \mathscr{M} = X = k' \cap \mathscr{M} = m' \cap k'$. Therefore $\ell' \cap k' = X$, and $\ell \cap k = \varnothing$. Since ℓ' and k' intersect, they are coplanar as projective lines, and ℓ and k are coplanar as well. Since ℓ and k do not intersect, they are parallel. ■

Note that the proof above implies that ℓ and m are parallel lines in $\mathscr{L}$ exactly when $\ell' \cap m'$ is in the maximal proper flat $\mathscr{M}$ in $(\mathscr{P}', \mathscr{L}')$.

It is possible to reverse the construction of Theorem 14 to give a second way to build a projective space from an affine one. Recall that a pencil of parallel lines is an equivalence class of lines with respect to the relation "is parallel or equal to." The corresponding definition for affine space is given below. Note that the following approach is analogous to the embedding of an affine plane in a projective plane described in Chapter 2, Theorem 13.

Definition. A *parallel class* of lines in an affine space $(\mathscr{P}, \mathscr{L}, \mathscr{E})$ is an equivalence class with respect to the relation "is parallel or equal to" on $\mathscr{L}$.

Note that a parallel class is a maximal set of mutually parallel lines. We next need a lemma concerning parallel lines and planes in Desarguesian affine spaces.

■ **LEMMA.** Let $(\mathscr{P}, \mathscr{L}, \mathscr{E})$ be a Desarguesian affine space, let ℓ and m be lines in a plane α that intersect at a point A, let B be a point not in α, and let ℓ' and m' be the lines through B parallel to ℓ and m, respectively. If h is any line in α containing A, and h' is the line through B parallel to h, then h' is in the plane β determined by ℓ' and m'.

Proof: Let X be any point on ℓ different from A, and let k be the line through X parallel to h. Then $k \in \alpha$, so $k \cap m = $ Y for some point Y. Let $X \in n \| \ell(A, B)$. Then n intersects ℓ' at a point X'. Similarly there is a point Y' in m' such that $\ell(Y, Y') \| \ell(A, B)$, as shown in Fig. 7.9. Using Desargues' Theorem on triangles A, X, Y and B, X', Y' gives $\ell(X, Y) \| \ell(X', Y')$, and it follows from the transitivity of parallelism that $\ell(X', Y') \| h'$. Thus $\ell(X', Y')$ and h' are coplanar, and since they must be in the plane determined by X', Y', and B, they are in β. ■

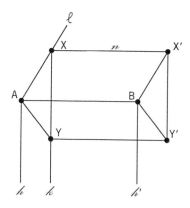

Figure 7.9

■   **THEOREM 15.** Let $(\mathscr{P}, \mathscr{L}, \mathscr{E})$ be an affine space. For each parallel class Γ of lines, let X_Γ be a new point adjoined to $\mathscr{P}$. Then $(\mathscr{P}', \mathscr{L}')$ is a projective space where

$$\mathscr{P}' = \mathscr{P} \cup \{X_\Gamma : \Gamma \text{ a parallel class of lines}\}.$$

For $\ell \in \mathscr{L}$, define ℓ' to be $\ell \cup \{X_\Gamma\}$ where ℓ is in the parallel class Γ. Define $\mathscr{M}$ to be $\{X_\Gamma : \Gamma \text{ a parallel class of lines}\}$. For each plane $\alpha \in \mathscr{E}$, let $\ell_\alpha = \{X_\Gamma : \Gamma \text{ contains a line in } \alpha\}$. Define $\mathscr{L}$ by $\mathscr{L}' = \{\ell' : \ell \in \mathscr{L}\} \cup \{\ell_\alpha : \alpha \in \mathscr{E}\}$.

Proof: If P and Q are in $\mathscr{P}$, then $\ell'(P, Q) = m'$, where $m = \ell(P, Q)$. Clearly the first axiom for projective space holds in this case. For P and X_Γ, the unique line containing them is k', where k is the (unique) line containing P in parallel class Γ.

For X_Γ and X_Ψ, take any line ℓ in Γ, and let A be a point on ℓ. Let m be the line in parallel class Ψ containing A. Then m and ℓ are intersecting affine lines, thus they lie in a unique plane α. Furthermore the unique line in $\mathscr{L}'$ containing X_Γ and X_Ψ is ℓ_α. Clearly ℓ_α contains the two points in question, and if any other line contains them, it must be of the form ℓ_β for some plane $\beta \in \mathscr{E}$. This means that there is a line ℓ' of parallel class Γ and a line m' of parallel class Ψ in β. These lines (being coplanar and in distinct parallel classes) intersect at a point B. Now if X_Ω is any point in ℓ_β, the line of Ω containing B is parallel to the line of Ω containing A. Since the former line is in β, the latter must be in α by the lemma. Thus X_Ω is in ℓ_α and $\ell_\beta \subseteq \ell_\alpha$. A similar argument gives the reverse inclusion, and the first axiom for projective space holds.

To prove the Pasch Axiom, note that if $\ell'(A, B)$ and $\ell'(C, D)$ have nonempty intersection (A, B, C, and D in $\mathscr{P}$), then the points A, B, C, and D are coplanar in $(\mathscr{P}, \mathscr{L}, \mathscr{E})$, whence $\ell(A, C)$ and $\ell(B, D)$ either intersect or are parallel. In the first instance they still intersect in $(\mathscr{P}', \mathscr{L}')$; in the second instance they intersect in X_Γ where Γ is their parallel class. If $\ell'(A, X_\Gamma) \cap \ell'(C, D) = E \in \mathscr{P}$, then A, C, D, and E were coplanar in $(\mathscr{P}, \mathscr{L}, \mathscr{E})$. Thus the line in the parallel class Γ that contains D intersected $\ell(A, C)$ at some point F. Consequently $\ell'(A, C) \cap \ell'(X_\Gamma, D) = F$. The proofs of the remaining cases are left as exercises.

To prove that the final axiom for projective space holds, note that each affine plane contains at least three parallel classes of lines; hence each line of the form ℓ_α contains at least three points. Since each affine line ℓ contains at least two points, each line of the form ℓ' contains at least three points. Since not all points are collinear in an affine space, not all points are collinear in the projective space. Further, if the affine space were nonplanar, the projective space would have dimension at least 3 as well.   ■

■ **THEOREM 16.** The $\mathcal{M}$ defined in the statement of Theorem 15 is a maximal proper flat of $(\mathcal{P}', \mathcal{L}')$.

Proof: It must be shown that $\overline{\mathcal{M} \cup \{A\}} = \mathcal{P}'$ for any A in $\mathcal{P}$. This in turn follows if it can be shown that any B in $\mathcal{P}$ is contained in $\overline{\mathcal{M} \cup \{A\}}$. But for such a B, $\ell(B, A)$ is in some parallel class Γ, so $B \in \ell'(X_\Gamma, A) \subseteq \overline{\mathcal{M} \cup \{A\}}$.   ■

Having now established the connections between affine and projective space, we will investigate the algebraic ramifications of these results in the next section by coordinatizing nonplanar projective spaces.

EXERCISES

1. Show that any line is a maximal proper flat of a projective plane and that any plane is a maximal proper flat of a projective 3-space.

2. Let $\mathcal{M}$ be a maximal proper flat in $(\mathcal{P}, \mathcal{L})$. Show that $\mathcal{P} = \overline{\{A\} \cup \mathcal{M}}$ for every point A not in $\mathcal{M}$.

3. Let $\mathcal{M}$ be a maximal proper flat of $(\mathcal{P}, \mathcal{L})$, and let α be a plane not contained in $\mathcal{M}$. Prove that α intersects $\mathcal{M}$ in exactly one line. [*Hint*: Let $\alpha = (P, \ell(A, B))$, and show that $\alpha \cap \mathcal{M} = \ell(C, D)$, where $C = \ell(P, A) \cap \mathcal{M}$ and $D = \ell(P, B) \cap \mathcal{M}$.]

4. Complete the proof Theorem 15 by proving the remaining cases of the Pasch Axiom.

5. Make an affine space out of the Fano space given in Example 2 by removing a plane. Coordinatize the resulting affine space.

6. Coordinatize one of the (projective) planes in the Fano space.

7. Construct a projective space from the affine space $AG(\mathbb{Z}_2^3)$ by the method given in Theorem 15.

*8. Let $F = GF(p^n)$, and consider the affine space $AG(F^3)$. Construct a projective space by adding a point to each parallel class of lines. What dimension is the resulting projective space? How many points and lines does it have?

7.5 COORDINATES IN PROJECTIVE SPACE

It will next be shown that any Desarguesian projective space $(\mathcal{P}', \mathcal{L}')$ can be coordinatized by a division ring $\mathbb{D}$ so that $\mathcal{P}'$ is the set of one-dimensional subspaces of a vector space $\mathcal{V}$ over $\mathbb{D}$ and $\mathcal{L}'$, the set of its two-dimensional subspaces. Such a projective space is denoted $PG(\mathcal{V})$. It will further be shown that the (projective) flats of $(\mathcal{P}', \mathcal{L}')$ are in 1-1 correspondence with the linear subspaces of $\mathcal{V}$. Since every nonplanar projective space is Desarguesian, the results will automatically hold for nonplanar projective spaces.

Let $(\mathcal{P}', \mathcal{L}')$ be a Desarguesian projective space, and let $\mathcal{M}$ be a maximal proper flat of $(\mathcal{P}', \mathcal{L}')$. Then $\mathcal{P}' \setminus \mathcal{M} = \mathcal{P}$ is an affine space, and therefore it can be coordinatized as a vector space over some division ring $\mathbb{D}$. Since $\mathbb{D}$ is a vector space over itself, $\mathscr{P} \times \mathbb{D} = \mathscr{V}$ is also a vector space over $\mathbb{D}$. The elements of $\mathscr{V}$ are ordered pairs of the form (P, a) where P is in $\mathscr{P}$ and a is an element of $\mathbb{D}$. The origin of $\mathscr{V}$ is $(O, 0)$.

The one-dimensional linear subspaces of a general (left) vector space $\mathscr{W}$ over $\mathbb{D}$ are the subsets of the form $\langle \mathbf{x} \rangle = \{m\mathbf{x} : m \in \mathbb{D}\}$, where $\mathbf{x}$ is a nonzero vector in $\mathscr{W}$. In the case where $\mathscr{W} = \mathscr{P} \times \mathbb{D}$, these subspaces can be written in one of the forms

$$\{m(P, 1) : m \in \mathbb{D}\} = \langle P, 1 \rangle,$$

or

$$\{m(P, 0) : m \in \mathbb{D}\} = \langle P, 0 \rangle,$$

where P is a point in $\mathscr{P}$.

Thus a 1-1 correspondence between the one-dimensional subspaces of $\mathscr{V}$ and the points of $\mathscr{P}'$ can be described as follows. If $P \in \mathscr{P}$, associate with P the one-dimensional subspace $\langle P, 1 \rangle$ of $\mathscr{V}$. If $Q \in \mathscr{M}$, then select any point P in $\ell'(O, Q)$ with P different from O and Q. Now P is not in $\mathscr{M}$ by Theorem 13. Thus $P \in \mathscr{P}$. Associate to Q the one-dimensional subspace $\langle P, 0 \rangle$ of $\mathscr{V}$. More formally, the following mapping can be defined:

Definition. Let f be the mapping of the points of the Desarguesian projective space $(\mathscr{P}', \mathscr{L}')$ to the one dimensional subspaces of $\mathscr{P} \times \mathbb{D}$ given by

$$f(P) = \langle P, 1 \rangle \qquad [P \in \mathscr{P}],$$
$$f(Q) = \langle P, 0 \rangle \qquad [Q = \mathscr{M} \cap \ell'(O, P)].$$

■ **THEOREM 18.** The mapping f is well defined, one-one, and onto.

Proof: For $P \in \mathscr{P}$, $f(P)$ is clearly well defined. If $Q = \mathscr{M} \cap \ell'(O, P) = \mathscr{M} \cap \ell'(O, S)$, then O, P, and S are collinear in $(\mathscr{P}', \mathscr{L}')$, and therefore are collinear in $\mathscr{P}$. Thus $S = tP$ for some t in $\mathbb{D}$ so that $\langle P, 0 \rangle = \langle tP, 0 \rangle = \langle S, 0 \rangle$, and likewise $f(Q)$ is well defined.

To show that f is one-one, note that if $\langle P, 1 \rangle = \langle Q, 1 \rangle$ in $\mathscr{P} \times \mathbb{D}$, then $P = Q$. Thus, if $f(P) = f(Q)$ for P and Q in $\mathscr{P}$, then $P = Q$. If $f(P) = f(Q) = \langle S, 0 \rangle$ for P and Q in $\mathscr{M}$, then $P = \ell'(O, S) \cap \mathscr{M} = Q$; once again, f is one-one. For $P \in \mathscr{P}$ and $Q \in \mathscr{M}$, $f(P) = \langle P, 1 \rangle$ and $f(Q) = \langle S, 0 \rangle$, so $f(P) \neq f(Q)$. Thus in all cases $f(P) = f(Q)$ implies that $P = Q$.

To show that f is onto, if $m = \langle S, b \rangle$ is a one-dimensional subspace of $\mathscr{P} \times \mathbb{D}$ with b different from 0, then $m = \langle b^{-1}S, 1 \rangle = f(b^{-1}S)$. If $b = 0$, then $m = f(Q)$, where $Q = \ell'(O, S) \cap \mathscr{M}$, so f is onto.   ■

A two-dimensional linear subspace $\mathscr{U}$ of a general (left) vector space $\mathscr{W}$ over $\mathbb{D}$ is an affine plane containing the origin of the affine space $\mathscr{W}$.

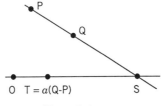

Figure 7.10

$\mathscr{U}$ is determined by any two of its vectors provided that they are noncollinear with the origin; alternatively, $\mathscr{U}$ is determined by any two of its one-dimensional linear subspaces. In case $\mathscr{W} = \mathscr{P} \times \mathbb{D}$, $\mathscr{U}$ is determined by any two of its nonzero subspaces $\langle P, a \rangle$ and $\langle Q, b \rangle$, where a and b can be chosen to be either 0 or 1.

■ **THEOREM 19.** Under the mapping f, the lines in $\mathscr{L}'$ are in one-one correspondence with the two-dimensional subspaces of $\mathscr{P} \times \mathbb{D}$.

Proof: Let $\ell'(P, Q)$ be a line in $\mathscr{L}'$. Then $f(P)$ and $f(Q)$ are two one-dimensional subspaces of $\mathscr{P} \times \mathbb{D}$. As such, they determine a unique affine plane, say, α, through the origin of $\mathscr{P} \times \mathbb{D}$. It can now be shown that $S \in \ell'(P, Q)$ if and only if $f(S) \subseteq \alpha$.

First, assume that S is in $\ell'(P, Q)$. If P, Q, and S are all in $\mathscr{P}$, then $S = P + t(Q - P)$ for some t in $\mathbb{D}$, and $(S, 1) = (P + t(Q - P), 1)$, which is clearly in α if $(P, 1)$ and $(Q, 1)$ are in α. If P and Q are in $\mathscr{P}$, while S is in $\mathscr{M}$, then $S = \ell'(P, Q) \cap \mathscr{M}$ and $f(S) = \ell((0, 0), (T, 0))$, where $T \in \ell'(O, S)$. Thus $\ell'(P, Q) \cap \ell'(O, T) = S \in \mathscr{M}$, so, back in the affine plane (when $\mathscr{M}$ is removed) we have $\ell(P, Q) \| \ell(O, T)$. Consequently $T = a(Q - P)$ for some a in $\mathbb{D}$ (see Fig. 7.10). Thus $(T, 0) = a(Q, 1) - a(P, 1)$ in $\mathscr{P} \times \mathbb{D}$, and $(T, 0)$ is in α when $(P, 1)$ and $(Q, 1)$ are in α. Therefore, in this case as well, if S is in $\ell'(P, Q)$, then $f(S)$ is in the two-dimensional linear subspace determined by $f(P)$ and $f(Q)$. If P is in $\mathscr{M}$ while Q and S are in $\mathscr{P}$, a similar argument shows that $f(S)$ is in α.

If two of P, Q, and S are in $\mathscr{M}$, then all three are in $\mathscr{M}$, where $\mathscr{M}$ is a flat of $(\mathscr{P}', \mathscr{L}')$. Taking P*, Q*, and S* as third points on $\ell'(O, P)$, $\ell'(O, Q)$, and $\ell'(O, S)$, respectively, then P*, Q* and S* are in the projective plane $(O, \ell'(P, Q))$, thus S* is in the (affine) plane determined by P*, Q*, and O in $\mathscr{P}$ (see Fig. 7.11). Therefore $S^* = tP^* + rQ^*$, so $(S^*, 0) = t(P^*, 0) + r(Q^*, 0)$. But the vectors on the right are in $f(P)$, and f(Q), respectively, so that $f(S) = \langle S^*, 0 \rangle$ in α which is the plane determined by $f(P)$ and $f(Q)$ in $\mathscr{V}$. The proof that if $f(S)$ is in α, then S is in $\ell'(P, Q)$ is left as an exercise.

Finally, it can be shown that the subspaces of $\mathscr{V}$ are in one-one correspondence with the flats of $(\mathscr{P}', \mathscr{L}')$. First, observe that $\mathscr{M}$ corresponds to the subspace $\mathscr{U} = \{(P, 0) : P \in \mathscr{P}\}$ of $\mathscr{V} = \mathscr{P} \times \mathbb{D}$.

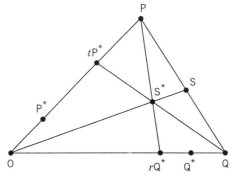

Figure 7.11

■ **THEOREM 20.** The flats of $(\mathscr{P}', \mathscr{L}')$ are in one-one correspondence with the linear subspaces of $\mathscr{V} = \mathscr{P} \times \mathbb{D}$ under the natural extension of f to a mapping of subsets of $\mathscr{P}'$, with $f(\varnothing)$ defined to be $\{(O, 0)\}$.

Proof: Let $\mathscr{S}$ be a flat of $(\mathscr{P}', \mathscr{L}')$. We first show that $f(\mathscr{S}) = \bigcup_{S \in \mathscr{S}} f(S)$ is a linear subspace of $\mathscr{V}$. Clearly $f(\mathscr{S})$ contains the origin of $\mathscr{V}$ for any $\mathscr{S}$ (including $\mathscr{S} = \varnothing$). Suppose that (P, a) and (Q, b) are in $f(\mathscr{S})$. It must be shown that $(tP + (1 - t)Q, ta + (1 - t)b) \in f(\mathscr{S})$.

Case 1. $a = b = 0$. Let $S = \ell'(O, P) \cap \mathscr{M}$, and $T = \ell'(O, Q) \cap \mathscr{M}$. Then $f(S) = \ell((O, 0), (P, 0))$ and $f(T) = \ell((O, 0), (Q, 0))$. Now $\ell'(O, tP + (1 - t)Q)) \cap \mathscr{M} = X$ for some X in $\ell'(S, T) \subseteq \mathscr{L}$. Thus $f(X) = \ell((O, 0), (tP + (1 - t)Q), 0) \subseteq f(\mathscr{S})$. (See Fig. 7.12.)

Case 2. $a \neq 0$, $b = 0$. Without loss of generality assume that $a = 1$ and $t \neq 0$. It must be shown that $(tP + (1 - t)Q, t)$ is in $f(\mathscr{S})$. But this is equivalent to showing that $(P + (t^{-1} - 1)Q, 1)$ is in $f(\mathscr{S})$, which is the case if $P + (t^{-1} - 1)Q$ is in $\mathscr{S}$. P is in $\mathscr{S}$, since $(P, 1)$ is in $f(\mathscr{S})$. Let $S = \ell'(O, P) \cap \mathscr{M}$, and let $T = \ell'(O, Q) \cap \mathscr{M} \subseteq \mathscr{S} \cap \mathscr{M}$. $P + (t^{-1} - 1)Q = \ell'(T, P) \cap \ell'(S, (t^{-1} - 1)Q)$, and since T and P are in $\mathscr{S}$, so is $P + (t^{-1} - 1)Q$. (See Fig. 7.13.)

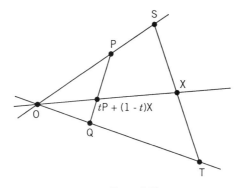

Figure 7.12

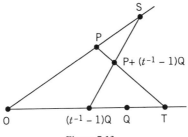

Figure 7.13

CASE 3. $a, b \neq 0$. Assume without loss of generality that $a = b = 1$. Then P and Q are in $\mathscr{S}$, so $t\mathrm{P} + (1 - t)\mathrm{Q} \in \mathscr{S}$ and $(t\mathrm{P} + (1 - t)\mathrm{Q}, 1) \in f(\mathscr{S})$.

To summarize, if $\mathscr{S}$ is a flat of $(\mathscr{P}', \mathscr{L}')$, then $f(\mathscr{S})$ is a vector subspace of $\mathscr{V}$.

To demonstrate the converse, let $\mathscr{U}$ be a subspace of $\mathscr{V}$. It must be shown that $\mathscr{U} = f(\mathscr{S})$ for some flat $\mathscr{S} \subseteq \mathscr{P}'$.

Let $\mathscr{S} = \{\mathrm{X} \in \mathscr{P}' : f(\mathrm{X}) \subseteq \mathscr{U}\}$. Let P and Q be points in $\mathscr{S}$ with $S \in \ell'(\mathrm{P}, \mathrm{Q})$. It must be shown that S is in $\mathscr{S}$, namely that $f(\mathrm{S}) \subseteq \mathscr{U}$. By definition of $\mathscr{S}$, $f(\mathrm{P})$ and $f(\mathrm{Q})$ are in $\mathscr{U}$. Suppose that $f(\mathrm{P}) = \langle \mathrm{R}, a \rangle$ and $f(\mathrm{Q}) = \langle \mathrm{T}, b \rangle$. Since $\mathscr{U}$ is a subspace of $\mathscr{V}$, $\{(\mathrm{O}, 0), (\mathrm{R}, a), (\mathrm{T}, b)\} \subseteq \mathscr{U}$. But $f(\mathrm{S}) \subseteq f(\ell'(\mathrm{P}, \mathrm{Q})) \subseteq \overline{\{(\mathrm{O}, 0), (\mathrm{R}, a), (\mathrm{T}, b)\}} \subseteq \mathscr{U}$.

The chapter ends with some examples of coordinatization.

□EXAMPLE 7: The Fano plane. Let $\mathscr{P}' = \{A, B, C, D, E, F, G\}$, and let $\mathscr{L}' = \{\{ABC\}, \{CDE\}, \{EFA\}, \{ADG\}, \{BGE\}, \{CGF\}, \{BDF\}\}$. Let $\mathscr{M} = \{BDF\}$; thus $\mathscr{P} = \{ACEG\}$ and $\mathscr{L} = \{AC\}, \{CE\}, \{EA\}, \{AG\}, \{GE\}, \{CG\}\}$. Thus $(\mathscr{P}, \mathscr{L})$ is the affine plane of order 2, and choosing $A = O$, the vector space $\mathscr{P}$ becomes the two-dimensional vector space over $\mathbb{Z}_2$, where C can be chosen as $(1, 0)$, E as $(0, 1)$ and G as $(1, 1)$. Hence $\mathscr{P} = \mathbb{Z}_2 \times \mathbb{Z}_2$ and $\mathscr{V} = \mathscr{P} \times \mathbb{Z}_2 = (\mathbb{Z}_2)^3$. The mapping f is given by

$f(A) = \langle 0, 0, 1 \rangle$
$f(C) = \langle 1, 0, 1 \rangle$
$f(E) = \langle 0, 1, 1 \rangle$
$f(G) = \langle 1, 1, 1 \rangle$
$f(B) = \langle 1, 0, 0 \rangle$, since $B \in \ell'(A, C)$ and the affine coordinates of C
$\qquad\qquad\qquad$ are $(1, 0)$
$f(D) = \langle 1, 1, 0 \rangle$
$f(F) = \langle 0, 1, 0 \rangle$

It follows that the Fano plane can be written as $P_2(\mathbb{Z}_2)$.   □

☐**EXAMPLE 8:** The final example is a 15-point, 35-line, 15-plane, projective space. Its planes are listed below for convenience, as well as its points and lines. The exercises that follow will lead the reader through the coordinatization of this space as $P_3(\mathbb{Z}_2)$.

$$\mathscr{P}' = \{ABCDEFGHIJKLMNO\},$$

$$\mathscr{L}' = \{\{ABI\}\{EFI\}\{GHI\}\{CDI\}\{ACJ\}\{EGJ\}\{FHJ\}\{ADK\}\{FGK\}$$
$$\{EHK\}\{BCK\}\{AEL\}\{BFL\}\{DHL\}\{CGL\}\{AFM\}\{GDM\}\{BEM\}$$
$$\{CHM\}\{AGN\}\{DFN\}\{BHN\}\{CEN\}\{AHO\}\{BGO\}\{DEO\}$$
$$\{CFO\}\{IML\}\{JLN\}\{IJK\}\{KMN\}\{JMO\}\{INO\}\{KLO\}\{BDJ\}\}.$$

The planes of the space are given by

$$\mathscr{E}' = \{\{ABEFIML\}\{BHDFNJL\}\{CDGHILM\}\{GHEFIJK\}\{BHCENKM\}$$
$$\{ABCDIJK\}\{AGDFNKM\}\{ACFHJMO\}\{ABGHINO\}\{BDEGJMO\}$$
$$\{ACEGJLN\}\{ADEHOKL\}\{CDEFINO\}\{BCGFKOL\}\{IJKLMNO\}\}.$$

☐

EXERCISES

Exercises 1–6 refer to the plane given in Example 8. Let O be as given, and let $\mathscr{M} = \{BHCENKM\}$.

1. List the points in $\mathscr{P}$, and make an addition table for them.
2. Show that the coordinatizing field is $\mathbb{Z}_2$, and make a scalar multiplication table for $\mathscr{P}$ as a $\mathbb{Z}_2$-vector space.
3. With O = (0,0,0), I = (1,0,0), and J = (0,1,0) and F = (0,0,1), coordinatize the remaining points of $\mathscr{P}$ as ordered triples from $\mathbb{Z}_2$, that is, as elements of $(\mathbb{Z}_2)^3$.
4. Coordinatize the points of $\mathscr{P}'$ as one-dimensional subspaces of $(\mathbb{Z}_2)^4$.
5. Coordinatize the lines of $\mathscr{L}'$ as two-dimensional subspaces of $(\mathbb{Z}_2)^4$.
6. To what three-dimensional linear subspace of $(\mathbb{Z}_2)^4$ does the projective plane {BCGFKOL} correspond? To what subspace does {GHEFIJK} correspond?
*7. State Pappus' Theorem for (Desarguesian) projective spaces, and show that $(\mathscr{P}', \mathscr{L}')$ satisfies it if and only if the coordinatizing division ring is a field. [*Hint*: Translate the statement of Pappus' Theorem for affine spaces, given in Section 3.7, into a projective space statement.]

SUGGESTED READING

H. S. M. Coxeter, *Projective Geometry*, 2d ed., New York: Springer-Verlag, 1987.

R. Hartshorne, *Foundations of Projective Geometry*, New York: Benjamin, 1967.

G. deB. Robinson, *The Foundations of Projective Space*, Toronto: Toronto University Press, 1940.

CHAPTER 8

Lattices of Flats

Just as (Desarguesian) affine and projective spaces can be presented either from the synthetic or the vector space point of view, they can also be studied through their flats. These flats will be analyzed here through closure spaces, and then through lattices.

8.1 CLOSURE SPACES

Recall that a closure space is a set $\mathscr{X}$, together with a *closure operator* or mapping $s : 2^{\mathscr{X}} \to 2^{\mathscr{X}}$ which is increasing, idempotent, and isotone (where $2^{\mathscr{X}}$ is the collection of subsets of $\mathscr{X}$). The basic examples discussed so far have been $(\mathscr{P}, \overline{})$, where $\mathscr{P}$ is the set of points of an affine space and $\overline{\mathscr{M}}$ is the *affine hull* of a subset $\mathscr{M}$ of $\mathscr{P}$ (i.e., the smallest affine flat containing $\mathscr{M}$), and $(\mathscr{P}', \overline{})$, where $\mathscr{P}'$ is the set of points of a projective space and $\overline{\mathscr{M}}$ is the smallest projective flat containing $\mathscr{M}$. A closure space $(\mathscr{P}, \overline{})$ will be called an *affine closure space*, while $(\mathscr{P}', \overline{})$ is a *projective closure space*. Some examples and properties of closure spaces are presented next.

□ **EXAMPLE 1:** Let $\mathscr{V}$ be a vector space over a division ring, and for $\mathscr{M} \subseteq \mathscr{V}$, define $s(\mathscr{M})$ to be the smallest linear subspace of $\mathscr{V}$ containing $\mathscr{M}$. Then $(\mathscr{V}, s)$ is a closure space. □

Definition. If $(\mathscr{X}, s)$ is a closure space, a subset $\mathscr{M}$ of $\mathscr{X}$ is said to be *closed* if $s(\mathscr{M}) = \mathscr{M}$.

The closed subsets of an affine closure space $(\mathscr{P}, \overline{})$ are the affine flats of $\mathscr{P}$, and the closed subsets of a projective closure space $(\mathscr{P}', \overline{})$ are the projective flats of $\mathscr{P}'$.

165

If $(\mathscr{X}, s)$ is any closure space, then $\mathscr{X}$ is closed because $\mathscr{X} \subseteq s(\mathscr{X})$ since s is increasing, but $s(\mathscr{X}) \subseteq \mathscr{X}$ since s is a mapping from the subsets of $\mathscr{X}$ to the subsets of $\mathscr{X}$. The empty set need not be closed; in Example 1 above $s(\varnothing)$ is the (nonempty) subspace consisting only of the zero vector.

Definition. If $(\mathscr{X}, s)$ is a closure space in which the empty set is closed, $(\mathscr{X}, s)$ is said to be *normalized*.

■ **THEOREM 1.** Arbitrary intersections of closed subsets of a closure space are closed.

Proof: Suppose that $\{\mathscr{M}_i\}_{i \in I}$ are closed subsets of the closure space $(\mathscr{X}, s)$. Then $\cap_{i \in I} \mathscr{M}_i \subseteq \mathscr{M}_i$ for each i, so $s(\cap_{i \in I} \mathscr{M}_i) \subseteq s(\mathscr{M}_i) = \mathscr{M}_i$ for each i; therefore $s(\cap_{i \in I} \mathscr{M}_i) \subseteq \cap_{i \in I} \mathscr{M}_i$. But the reverse inclusion follows from the fact that s is increasing; thus $s(\cap_{i \in I} \mathscr{M}_i) = \cap_{i \in I} \mathscr{M}_i$ and $\cap_{i \in I} \mathscr{M}_i$ is closed.   ■

Corollary. If $\mathscr{M}$ and $\mathscr{N}$ are closed subsets of the closure space $(\mathscr{X}, s)$, then $\mathscr{M} \cap \mathscr{N}$ is also closed.

■ **THEOREM 2.** Suppose that $\mathscr{J}$ is a collection of subsets of a set $\mathscr{X}$ which is closed under arbitrary intersections. Then $(\mathscr{X}, s)$ is a closure space where $s(\mathscr{M}) = \cap \{\mathscr{C} : \mathscr{C} \in \mathscr{J} \text{ and } \mathscr{M} \subseteq \mathscr{C}\}$.

Proof: Since $\mathscr{M}$ is a subset of each of the sets $\mathscr{C}$, $\mathscr{M}$ is a subset of their intersection; therefore $\mathscr{M} \subseteq s(\mathscr{M})$ and s is increasing. Since $s(\mathscr{M})$ is in $\mathscr{J}$, $s(\mathscr{M})$ is one of the sets being intersected to form $s(s(\mathscr{M}))$, and therefore $s(s(\mathscr{M})) \subseteq s(\mathscr{M})$. But since s is increasing, the reverse inclusion holds, so s is idempotent. If $\mathscr{M} \subseteq \mathscr{N}$, then any set containing $\mathscr{N}$ contains $\mathscr{M}$. In particular, $\mathscr{M} \subseteq s(\mathscr{N})$, so $s(\mathscr{M}) \subseteq s(\mathscr{N})$ and s is isotone.   ■

□**EXAMPLE 2:** Suppose that $(\mathscr{G}, \cdot)$ is a group. For $\mathscr{M} \subseteq \mathscr{G}$, define $C(\mathscr{M})$ to be $\{g \in \mathscr{G} : gm = mg \text{ for all } m \text{ in } \mathscr{M}\}$. $C(\mathscr{M})$ is called the *centralizer of* $\mathscr{M}$. $(\mathscr{G}, s)$ is a closure space where $s(\mathscr{M}) = C(C(\mathscr{M}))$.   □

□**EXAMPLE 3:** Suppose that K is a field and F is a subfield of K. For $\mathscr{M} \subseteq K$ define $s(\mathscr{M})$ to be the smallest subfield of K containing $\mathscr{M} \cup F$. Then (K, s) is a closure space.   □

□**EXAMPLE 4:** Let $\mathscr{X}$ be any nonempty set, and let $f : \mathscr{X} \to \mathscr{X}$ be a function on $\mathscr{X}$. Then s is a closure operator on $\mathscr{X}$ where $s(\mathscr{M}) = f^{-1}(f(\mathscr{M}))$.   □

□**EXAMPLE 5:** If $\mathscr{X}$ is a topological space and $c(\mathscr{M})$ is defined to be the topological closure of the subset $\mathscr{M} \subseteq \mathscr{X}$, then $(\mathscr{X}, c)$ is a closure space. In this example the *union* as well as the intersection of any two closed sets is closed.   □

□**EXAMPLE 6:** For $\mathcal{M}$ a subset of the Euclidean space $\mathbb{R}^n$, let $c(\mathcal{M})$ be the convex hull of $\mathcal{M}$, that is, the intersection of all convex subsets of $\mathbb{R}^n$ containing $\mathcal{M}$ as a subset. Then $(\mathbb{R}^n, c)$ is a closure space whose closed sets are the convex subsets of $\mathbb{R}^n$. (Recall that a subset $\mathcal{C} \subseteq \mathbb{R}^n$ is *convex* if, whenever $\mathcal{C}$ contains the points **x** and **y**, it also contains the *segment* they determine, namely $\{a\mathbf{x} + (1 - a)\mathbf{y} : 0 \leq a \leq 1\}$).   □

EXERCISES

1. Suppose that $\mathcal{V}$ is a vector space over a division ring $\mathbb{D}$, and for $\mathcal{M} \subseteq \mathcal{V}$, $s(\mathcal{M})$ is the smallest linear subspace of $\mathcal{V}$ containing $\mathcal{M}$.

   a. Show that $s(\mathcal{M})$ consists of all finite linear combinations of vectors in $\mathcal{M}$.

   b. If $\mathcal{W}$ and $\mathcal{U}$ are linear subspaces of $\mathcal{V}$, show that $s(\mathcal{W} \cup \mathcal{U}) = \{\mathbf{w} + \mathbf{u} : \mathbf{w} \in \mathcal{W}, \mathbf{u} \in \mathcal{U}\}$.

   c. Suppose that $\mathcal{T}$, $\mathcal{W}$, and $\mathcal{U}$ are subspaces of $\mathcal{V}$ with $\mathcal{T} \subseteq \mathcal{U}$. Show that $s(\mathcal{T} \cup (\mathcal{W} \cap \mathcal{U})) = s(\mathcal{T} \cup \mathcal{W}) \cap \mathcal{U}$.

2. Prove that $(\mathcal{G}, s)$ of Example 2 is a closure space.

3. Prove that the closed subsets of the closure space defined in Theorem 2 are exactly the sets in $\mathcal{J}$.

4. Use Theorem 2 to prove that (K, s), given in Example 3, is a closure space.

5. Let $(\mathcal{X}, s)$ be a closure space. Prove that $\mathcal{M}$ is a closed subset of $\mathcal{X}$ if and only if there is a subset $\mathcal{N} \subseteq \mathcal{X}$ such that $s(\mathcal{N}) = \mathcal{M}$.

6. Prove that $(\mathcal{X}, s)$ given in Example 4 is a closure space.

7. In Example 6, show that for $\mathcal{M}$ and $\mathcal{N}$ nonempty convex subsets of $\mathbb{R}^n$, $c(\mathcal{M} \cup \mathcal{N}) = \{a\mathbf{x} + (1 - a)\mathbf{y} : \mathbf{x} \in \mathcal{M}, \mathbf{y} \in \mathcal{N}, 0 \leq a \leq 1\}$.

8. Which of the Examples 2–6 are *normalized* closure spaces?

8.2 SOME PROPERTIES OF CLOSURE SPACES

It is possible to characterize projective spaces as closure spaces with several further properties, and it is these properties that will be defined in this section.

   Definition. A closure space $(\mathcal{X}, s)$ is said to be *point-closed* if whenever $x \in \mathcal{X}$, $s(\{x\}) = \{x\}$. Unless confusion threatens, we will use the notation $s(x)$ for the closure of the singleton subset $\{x\} \subseteq \mathcal{X}$, instead of the clumsier $s(\{x\})$.

   Affine and projective closure spaces are always point-closed, since single points are flats in each case.

   If $\mathcal{G}$ is a group with identity e, and for $\mathcal{M} \subseteq \mathcal{G}$, $s(\mathcal{M})$ is the smallest subgroup of $\mathcal{G}$ containing $\mathcal{M}$, $(\mathcal{G}, s)$ is not point-closed unless $\mathcal{G}$ is the trivial one-element group, since, given $x \neq e$, $s(x)$ must at least contain e as well as x.

Definition. A closure space $(\mathcal{X}, s)$ is *locally finite* if, whenever $\mathcal{M} \subseteq \mathcal{X}$ and $x \in s(\mathcal{M})$, there is a finite subset $\mathcal{F} \subseteq \mathcal{M}$ such that $x \in s(\mathcal{F})$.

■   **THEOREM 3.** Every projective closure space is locally finite.

Proof: Suppose that $(\mathcal{P}', \mathcal{L}')$ is a projective space and that $(\mathcal{P}', \overline{})$ is the corresponding closure space. Suppose that $\mathcal{M} \subseteq \mathcal{P}'$. Define

$$\mathcal{U} = \cup \{\overline{\mathcal{F}} : \mathcal{F} \text{ a finite subset of } \mathcal{M}\}.$$

Then $\mathcal{M} \subseteq \mathcal{U} \subseteq \overline{\mathcal{M}}$. Thus if it can be shown that $\mathcal{U}$ is closed (i.e., a flat), $\mathcal{U}$ will be equal to $\overline{\mathcal{M}}$. Then, if a point P is in $\overline{\mathcal{M}}$, P is in $\overline{\mathcal{F}}$ for some finite subset $\mathcal{F}$ of $\mathcal{M}$, and the closure space is locally finite.

Suppose that A and B are in $\mathcal{U}$ with $C \in \ell(A, B)$. Then $A \in \overline{\mathcal{F}}$ and $B \in \overline{\mathcal{G}}$ where $\mathcal{F}$ and $\mathcal{G}$ are finite subsets of $\mathcal{M}$. Therefore $\ell(A, B) \subseteq \overline{(\mathcal{F} \cup \mathcal{G})}$, and since $\mathcal{F} \cup \mathcal{G}$ is finite, C is in the closure of a finite subset of $\mathcal{M}$, thus C is in $\mathcal{U}$, and $\mathcal{U}$ is closed.   ■

Definition. A closure space $(\mathcal{X}, s)$ is called *additive* if whenever $\mathcal{M}$ and $\mathcal{N}$ are nonempty closed subsets of $\mathcal{X}$,

$$s(\mathcal{M} \cup \mathcal{N}) = \bigcup_{\substack{m \in \mathcal{M} \\ n \in \mathcal{N}}} s(\{m, n\}).$$

■   **THEOREM 4.** Every projective closure space is additive.

Proof: Recall (Chapter 7, Theorem 5) that if $\mathcal{M}$ is a flat of a projective space and Q is a point not in $\mathcal{M}$, then $\overline{\{Q\} \cup \mathcal{M}} = \cup_{A \in \mathcal{M}} \ell(Q, A)$; thus the result holds if one of $\mathcal{M}$ or $\mathcal{N}$ is a singleton. If $\mathcal{M}$ and $\mathcal{N}$ are arbitrary flats, then

$$\mathcal{M} \cup \mathcal{N} \subseteq \bigcup_{\substack{A \in \mathcal{M} \\ B \in \mathcal{N}}} \overline{\{A, B\}} \subseteq \overline{\mathcal{M} \cup \mathcal{N}}$$

(where $\overline{\{A, B\}}$ is simply $\ell(A, B)$). Since $\overline{\mathcal{M} \cup \mathcal{N}}$ is the smallest flat containing $\mathcal{M} \cup \mathcal{N}$, if the middle set (which will be referred to as $\mathcal{U}$) is closed (i.e., a flat), the theorem is proved.

Let $D, E \in \mathcal{U}$, with $F \in \ell(D, E)$. Then there are points A_1, A_2 in $\mathcal{M}$ and B_1, B_2 in $\mathcal{N}$, with $D \in \ell(A_1, B_1) = \overline{\{A_1, B_1\}}$ and $E \in \ell(A_2, B_2) = \overline{\{A_2, B_2\}}$. Thus $F \in \overline{\{D, E\}} \subseteq \overline{\{A_1, B_1, A_2, B_2\}} = \{A_1\} \cup \overline{\{A_2, B_1, B_2\}}$, so $F \in \ell(A_1, C)$, where $C \in \overline{\{A_2, B_1, B_2\}}$. But this implies that $C \in \ell(A_2, B_3)$, where $B_3 \in \ell(B_1, B_2) \subseteq \mathcal{N}$. Thus $F \in \ell(A_1, C) \subseteq \overline{\{A_1, A_2, B_3\}}$, so $F \in \ell(A_3, B_3)$, where $A_3 \in \ell(A_1, A_2) \subseteq \mathcal{M}$. Thus F is in $\mathcal{U}$, and $\mathcal{U}$ is closed.   ■

Definition. A closure space $(\mathcal{X}, s)$ is said to satisfy the *Steinitz-Mac Lane exchange property* if whenever $x \in s(\mathcal{M} \cup \{y\})$, and $x \notin s(\mathcal{M})$, then $y \in s(\mathcal{M} \cup \{x\})$.

■ **THEOREM 5.** For a closure space $(\mathcal{X}, s)$ that satisfies the Steinitz-Mac Lane exchange property, if $x \in s(\mathcal{M} \cup \{y\})$ and $x \notin s(\mathcal{M})$, then $s(\mathcal{M} \cup \{x\}) = s(\mathcal{M} \cup \{y\})$.

Proof: Since $\mathcal{M} \cup \{x\} \subseteq s(\mathcal{M} \cup \{y\})$, $s(\mathcal{M} \cup \{x\}) \subseteq s(\mathcal{M} \cup \{y\})$. Conversely, by the exchange property, $\mathcal{M} \cup \{y\} \subseteq s(\mathcal{M} \cup \{x\})$, so the reverse inequality holds. ■

■ **THEOREM 6.** Every projective closure space satisfies the Steinitz-Mac Lane exchange property.

Proof: The proof follows immediately from the Exchange Theorem (Chapter 6, Theorem 7). ■

The final definition needed to characterize projective spaces in terms of closure operators is irreducibility.

Definition. A closure space $(\mathcal{X}, s)$ is *irreducible* if whenever x, y are distinct elements in $\mathcal{X}$ there is a third element $z \neq x, y$ with $z \in s(\{x, y\})$.

■ **THEOREM 7.** Every projective closure space is irreducible.

Proof: This is simply a closure space statement of the axiom that each line contains at least three points. ■

EXERCISES

1. Prove that if $|\mathcal{X}| \geq 2$, then any closure space $(\mathcal{X}, s)$ is normalized if it is point-closed. [*Hint*: Show the contrapositive.]
2. Find an example of a closure space that is normalized but not point-closed.
3. Prove that every affine closure space is locally finite and has the Steinitz-Mac Lane exchange property.
4. Prove that for $\mathcal{V}$ a vector space and $s(\mathcal{M})$ the linear span of $\mathcal{M}$ (the smallest linear subspace of $\mathcal{V}$ containing $\mathcal{M}$), $(\mathcal{V}, s)$ is locally finite and additive.
5. Show that the closure space $(\mathbb{R}^n, c)$ whose closed subsets are the convex subsets of $\mathbb{R}^n$ (Example 6 above) is additive and locally finite. Show further that it does not satisfy the Steinitz-Mac Lane exchange property.
6. Show that no affine closure space is additive.
7. Prove that for $(\mathcal{X}, s)$ a closure space with $\mathcal{M}$ and $\mathcal{N}$ subsets of $\mathcal{X}$, that $\overline{\mathcal{M} \cup \mathcal{N}} \subseteq \overline{\mathcal{M}} \cup \overline{\mathcal{N}}$. Show, by giving an example, that the reverse inclusion is not necessarily true.

8.3 PROJECTIVE CLOSURE SPACES

In the preceding section, six properties of projective closure spaces were defined. This section will show that conversely, any closure space that has these properties is a projective closure space.

■ **THEOREM 8.** Suppose that $(\mathcal{X}, s)$ is a normalized, point-closed, locally finite, additive, irreducible closure space with the Steinitz-Mac Lane exchange property. Then $(\mathcal{P}, \mathcal{L})$ is a projective space where the points are the one-element subsets of $\mathcal{X}$ (i.e., $\mathcal{P} = \mathcal{X}$) and the lines are the closures of the two-element subsets of $\mathcal{X}$ (i.e., $\ell \in \mathcal{L}$ if $\ell = s(\{X, Y\})$ for some $X \neq Y$ in $\mathcal{P}$). Furthermore the closed subspaces of $(\mathcal{P}, s)$ are exactly the flats of $(\mathcal{P}, \mathcal{L})$.

Proof: PS1. Given $X \neq Y$ in $\mathcal{P}$, they are both in the line $s(\{X, Y\})$. Say they are also in the line $s(\{P, Q\})$. Then if $X \neq P$, $X \notin s(P)$, since $s(P) = P$. By the exchange property, since $X \in s(\{P, Q\})$ and $X \notin s(P)$, then $Q \in s(\{X, P\})$. Thus $\{P, Q\} \subseteq s(\{X, P\})$, so $s(\{P, Q\}) \subseteq s(\{X, P\})$, and in particular $Y \in s(\{X, P\})$. If $Y \neq P$, then, again, applying the exchange property gives $P \in s(\{X, Y\})$. Using the same argument with Q replacing P gives $Q \in s(\{X, Y\})$. Thus $s(\{P, Q\}) \subseteq s(\{X, Y\})$. Since the reverse inclusion holds trivially, the two lines are equal.

PS2. Suppose that $s(\{A, B\}) \cap s(\{C, D\}) = E$ for some $E \in \mathcal{X}$. Then, assuming that all five points are distinct, the exchange property gives $A \in s(\{B, E\}) \subseteq s(\{B, C, D\}) = s(\{C\} \cup s(\{B, D\}))$, and by additivity there is an F in $s(\{B, D\})$ such that $A \in s(\{C, F\})$. Using the exchange property again, $F \in s(\{A, C\})$.

PS3. There are at least three points on each line by irreducibility. If there is only one line, the space is called a *projective line*; if there is more than one line, it is a projective space in the usual sense.

The only properties not used so far are normalized and locally finite. These properties are needed to show that the closed subspaces of $(\mathcal{X}, s)$ are exactly the flats of $(\mathcal{P}, \mathcal{L})$. Suppose that $\mathcal{M}$ is a closed subspace of $(\mathcal{X}, s)$. If $\mathcal{M}$ is the empty set or a single point, then $\mathcal{M}$ is a flat of $(\mathcal{P}, \mathcal{L})$. If $\mathcal{M}$ contains two distinct points X and Y, then $\ell(X, Y) = s(\{X, Y\}) \subseteq s(\mathcal{M}) = \mathcal{M}$, and $\mathcal{M}$ is a flat.

Suppose that $\mathcal{M}$ is a flat of $(\mathcal{P}, \mathcal{L})$. If $\mathcal{M} = \varnothing$ or $\mathcal{M}$ is a single point, then $\mathcal{M}$ is closed since $(\mathcal{X}, s)$ is normalized and point-closed. Now assume that $\mathcal{M}$ has at least two distinct points. It must be shown that $\mathcal{M} = s(\mathcal{M})$. Suppose that X is any point in $s(\mathcal{M})$. Then $X \in s(\{A_1, \ldots, A_k\})$ where the A_i are in $\mathcal{M}$ since $(\mathcal{X}, s)$ is locally finite. If $k = 1$ or 2, then $X \in \mathcal{M}$. Assume that for $k = 1, 2, \ldots, n - 1$, $X \in \mathcal{M}$. If $X \in s(\{A_1, \ldots, A_n\})$, by additivity there exists $B \in s(\{A_1, \ldots, A_{n-1}\}) \subseteq \mathcal{M}$ such that $X \in s(\{B, A_n\})$. By the induction hypothesis, $B \in \mathcal{M}$, so $X \in \mathcal{M}$ and $\mathcal{M} = s(\mathcal{M})$. ■

Corollary. If $(\mathscr{X}, s)$ is a point-closed, additive, irreducible closure space with the Steinitz-Mac Lane exchange property, then $(\mathscr{P}, \mathscr{L})$ is a projective space where $\mathscr{P} = \mathscr{X}^{\cdot}$ and the lines are the closures of the two-element subsets of $\mathscr{X}$.

A projective space can now be viewed from three perspectives: *synthetically* as a set of points and a collection of distinguished subsets (lines) satisfying three axioms, *algebraically* as the subspaces of a vector space over a division ring, and as a *closure space* with six properties.

A final point of view is the *lattice theoretic* approach to projective space, first investigated by Garrett Birkhoff and Karl Menger in the 1930s, which will be described next. Here the flats of a projective space, together with the relation $\subseteq$ between the flats, is studied.

8.4 INTRODUCTION TO LATTICES

In the definitions that follow, the reader should bear in mind the example of the flats of a projective space, with the inclusion relation $\subseteq$.

Definition. A *partially ordered set* (or *poset*) is a pair $(S, \le)$ where S is a set and $\le$ a binary relation satisfying the following conditions.

P1. $x \le x$ for all x in S ($\le$ is reflexive).

P2. $x \le y$ and $y \le x$ imply that $x = y$ ($\le$ is antisymmetric).

P3. $x \le y$ and $y \le z$ imply that $x \le z$ ($\le$ is transitive).

Definition. A partially ordered set $(L, \le)$ is a *lattice* if every pair of its elements has a least upper bound and a greatest lower bound. Given x and y in L, z is their *least upper bound* if $x \le z$, $y \le z$, and whenever $x, y \le t$, then $z \le t$. For x and y, w is their *greatest lower bound* if $w \le x$, $w \le y$, and whenever $u \le x, y$, then $u \le w$.

The least upper bound of x and y in a lattice L is written $x \vee y$ (read "x join y") and their greatest lower bound is written $x \wedge y$ (read "x meet y").

□**EXAMPLE 7:** If $\mathscr{X}$ is a set, then $(2^{\mathscr{X}}, \subseteq)$ is a lattice where $A \vee B = A \cup B$ and $A \wedge B = A \cap B$.   □

□**EXAMPLE 8:** The subgroups of a group form a lattice, whose order relation is again set inclusion. The meet operation is once again intersection; however, the union of two subgroups is not necessarily a subgroup (see Exercise 7). Thus for H and K subgroups of G, $H \vee K$ is the smallest subgroup of G containing H and K.   □

□**EXAMPLE 9:** The linear subspaces of any vector space $\mathscr{V}$ form a lattice, again under set inclusion and with meets as intersections. The join of two subspaces $\mathscr{U}$ and $\mathscr{W}$ is their algebraic sum $\mathscr{U} + \mathscr{W} = \{\mathbf{u} + \mathbf{w} : \mathbf{u} \in \mathscr{U}, \mathbf{w} \in \mathscr{W}\}$.   □

In the next example, the order relation is not set inclusion.

□**EXAMPLE 10:** Let $\mathbb{N}$ be the set of natural numbers $1, 2, 3, \ldots$. For a, b in $\mathbb{N}$, define $a|b$ to mean that b is a multiple of a. Then $(\mathbb{N}, |)$ is easily verified to be a poset. The join of two natural numbers a and b is c iff $a|c$, $b|c$, and if a and b both divide d, then c divides d. Thus $a \vee b$ is the least common multiple of a and b. A similar argument shows that $a \wedge b$ is the greatest common divisor of a and b.   □

Many lattices, including those in Examples 8 and 9, are described in the following example.

□**EXAMPLE 11:** Suppose that $(\mathscr{X}, s)$ is a closure space. Then its closed subspaces form a lattice $\mathscr{C}(\mathscr{X}, s)$ under set inclusion. The meet of two closed subsets is their intersection (see the corollary to Theorem 1 above), and if $\mathscr{M}$ and $\mathscr{N}$ are closed subspaces, their join $\mathscr{M} \vee \mathscr{N}$ is given by $s(\mathscr{M} \cup \mathscr{N})$.   □

In view of Example 11, the flats of both affine and projective spaces are lattices, namely $\mathscr{C}(\mathscr{P}, \overline{})$, where $(\mathscr{P}, \mathscr{L}, \mathscr{E})$ is an affine space, and $\mathscr{C}(\mathscr{P}', \overline{})$, where $(\mathscr{P}', \mathscr{L}')$ is a projective space. In what follows the properties of these lattices will be discussed, and the properties characterizing lattices of flats of projective spaces will be developed. For a lattice theoretic characterization of the flats of an affine space; see Bennett (1983). First, however, an alternative definition of a lattice is presented.

Besides being partially ordered sets with additional properties, lattices can be defined as algebraic systems, using join and meet as binary operations defined in terms of axioms.

Alternative Definition of Lattices. A *lattice* is a triple $(L, \vee, \wedge)$ where L is a nonempty set, and $\vee$ and $\wedge$ are binary operations on L satisfying the following axioms for all x, y, z in L:

$L1.$ $x \vee x = x$; $x \wedge x = x$ (idempotent).

$L2.$ $x \vee y = y \vee x$; $x \wedge y = y \wedge x$ (commutative).

$L3.$ $x \vee (y \vee z) = (x \vee y) \vee z$; $x \wedge (y \wedge z) = (x \wedge y) \wedge z$ (associative).

$L4.$ $x \vee (x \wedge y) = x$; $x \wedge (x \vee y) = x$ (absorptive).

If $(L, \leq)$ is a partially ordered set in which every pair of elements has a join and a meet, it is easy to verify that $L1$, $L2$, $L3$, and $L4$ hold for those operations.

■ **LEMMA.** If x and y are elements of the lattice $(L, \vee, \wedge)$, then $x \vee y = y$ if and only if $x \wedge y = x$.

Proof: If $x \vee y = y$, then $x \wedge y = x \wedge (x \vee y) = x$ (by $L4$), and conversely, if $x \wedge y = x$, then $x \vee y = (x \wedge y) \vee x = x$ (by $L4$). Thus the conditions $x \vee y = y$ and $x \wedge y = x$ are equivalent. ■

■ **THEOREM 9.** If $(L, \vee, \wedge)$ satisfies $L1, \ldots, L4$, and if $x \leq y$ is defined to mean $x \vee y = y$, then $(L, \leq)$ is a lattice. Furthermore the operations $\vee$ and $\wedge$ serve as the least upper bound and the greatest lower bound, respectively.

Proof: First, $x \leq x$, since $x \vee x = x$ by $L1$; thus $\leq$ is reflexive. If $x \leq y$ and $y \leq x$, $y = x \vee y = y \vee x = x$, so $\leq$ is antisymmetric. If $x \leq y$ and $y \leq z$, then $z = z \vee y = z \vee (x \vee y) = z \vee (y \vee x) = (z \vee y) \vee x = z \vee x$, so $x \leq z$ and $\leq$ is transitive. Thus $(L, \leq)$ is a poset.

To show that $\vee$ and $\wedge$ are effective as the least upper bound and greatest lower bound, respectively, first note that $x \vee (x \vee y) = (x \vee x) \vee y = x \vee y$, so $x \leq x \vee y$. A similar argument shows that $y \leq x \vee y$. If x and y are both $\leq t$, then $t = t \vee x = t \vee y$. Thus $t = t \vee t = (t \vee x) \vee (t \vee y) = (t \vee t) \vee (x \vee y) = t \vee (x \vee y)$, and $x \vee y \leq t$. Therefore $x \vee y$ is the least upper bound of x and y.

By the lemma, $x \leq y$ if and only if $x \wedge y = x$. Using this definition, a similar argument can be used to show that $x \wedge y$ is the greatest lower bound for x and y. The proof is left as an exercise. ■

From now on we will use the two definitions of lattices interchangeably. So far the operations $\vee$ and $\wedge$ have been defined only for pairs of elements, although by the associative laws $(L3)$ arbitrary finite joins and meets can be defined as well. In general, abstract algebra deals only with *finitary* operations (ring addition, group multiplication, etc.); however, lattices are an exception because in some cases joins and meets of infinite sets of elements exist.

Definition. If $(L, \vee, \wedge)$ is a lattice, and $\{x_i\}_{i \in I}$ is a subset of L, $\bigvee_{i \in I} x_i$ is defined and equals z if

1. $x_i \leq z$ for all $i \in I$,
2. if $x_i \leq t$ $(i \in I)$ for some t, then $z \leq t$.

Similarly $\bigwedge_{i \in I} x_i$ is defined and equals w if

    1'. $w \le x_i$ for all $i \in I$,

    2'. if $y \le x_i$ $(i \in I)$ for some y, then $y \le w$.

Definition. A lattice is called *complete* if each of its subsets has a join and a meet.

If $\mathscr{S}$ is an arbitrary subset of a complete lattice L, we will refer to its join as $\bigvee \mathscr{S}$ and its meet as $\bigwedge \mathscr{S}$.

The closure space lattices $\mathscr{C}(\mathscr{X}, s)$ are always complete, but the division lattice $(\mathbb{N}, |)$ is not, since one cannot calculate least common multiples for arbitrary infinite collections of natural numbers. The lattice $2^{\mathscr{X}}$ of *all* subsets of a set $\mathscr{X}$ is complete, but the lattice of all *finite* subsets of an infinite set $\mathscr{X}$ is not.

EXERCISES

1. Verify that $(\mathbb{N}, |)$, as given in Example 10, is a lattice in which $a \wedge b$ is the greatest common divisor of the natural numbers a and b.

2. Verify that for $(\mathscr{X}, s)$ a closure space, the join $\mathscr{M} \vee \mathscr{N}$ in $\mathscr{C}(\mathscr{X}, s)$ (as defined in Example 11) is given by $s(\mathscr{M} \cup \mathscr{N})$.

3. Let $(L, \le)$ be a lattice. Show that the operations $\vee$ and $\wedge$ are commutative, associative, and idempotent.

4. a. Show that the real numbers, with their usual order, form a lattice. Describe the join and the meet of two arbitrary real numbers in this lattice.

    b. Show, by giving examples, that if $\mathscr{S}$ is an infinite subset of $\mathbb{R}$, $\bigvee \mathscr{S}$ and $\bigwedge \mathscr{S}$ might not exist.

    c. Give an example of an infinite subset $\mathscr{S}$ of $\mathbb{R}$ such that $\bigvee \mathscr{S}$ exists but is not in $\mathscr{S}$. Do the same for $\bigwedge \mathscr{S}$. Can this happen if $\mathscr{S}$ is finite?

5. Show that the finite subsets of an infinite set $\mathscr{X}$ form a lattice under set inclusion. Show that the finite and cofinite subsets of $\mathscr{X}$ also form a lattice under set inclusion. Is either lattice complete? (A subset $\mathscr{S} \subseteq \mathscr{X}$ is *cofinite* iff its complement $\mathscr{X} \setminus \mathscr{S}$ is finite.)

6. Complete the proof of Theorem 9 by showing that if $x \le y$ is defined to mean $x \wedge y = x$ in a lattice $(L, \vee, \wedge)$, then $x \wedge y$ is the greatest lower bound of x and y in $(L, \le)$.

7. For H and K subgroups of a group G, show that $H \cup K$ is also a subgroup if and only if $H \subseteq K$ or $K \subseteq H$.

8. Show that $L1$ is a consequence of $L2$, $L3$, and $L4$.

8.5 BOUNDED LATTICES: DUALITY

If a lattice has a largest element, it is denoted 1; if it has a smallest element, it is called 0. The elements $\{e\}$ and $\mathscr{G}$ are the 0 and 1 of the lattice of subgroups of the group $\mathscr{G}$ (whose identity element is e). For the lattice of flats of the projective space $(\mathscr{P}', \mathscr{L}')$, the 0 element is $\varnothing$, and the 1 element is $\mathscr{P}'$. A lattice that has both a 0 and a 1 is called *bounded*.

■ **THEOREM 10.** Every complete lattice is bounded.

Proof: The 0 element of a complete lattice L is clearly $\bigwedge\{x : x \in L\}$, which exists since L is complete; similarly the largest element of L is the join of all the elements of L.   ■

■ **THEOREM 11.** Suppose that L is a bounded lattice. The following conditions are equivalent:

   i. Every subset of L has a meet.
   ii. Every subset of L has a join.
   iii. L is complete.

Proof: We will show that i implies ii. The proof that ii implies i is left as an exercise. Since iii trivially implies i and ii, this will complete the proof of the theorem.

i → ii: Suppose that $\mathscr{S}$ is a subset of L. Let $\mathscr{T} = \{x \in L : s \leq x$ for all $s \in \mathscr{S}\}$. Now $\mathscr{T}$ is not empty, since $1 \in \mathscr{T}$. By hypothesis, $\mathscr{T}$ has a meet, say, t. For any $s \in \mathscr{S}$, since $s \leq x$ for all $x \in \mathscr{T}$, s is a lower bound of $\mathscr{T}$. It follows that s is under the greatest lower bound of $\mathscr{T}$, so $s \leq t$, and t is an upper bound for $\mathscr{S}$. If y is any other upper bound of $\mathscr{S}$, y is in $\mathscr{T}$ by definition. But t is the meet of $\mathscr{T}$, so $t \leq y$. Therefore t is the least upper bound, or join, of $\mathscr{S}$.   ■

Corollary. If $(\mathscr{X}, s)$ is a closure space, its lattice of closed subspaces, $\mathscr{C}(\mathscr{X}, s)$ is bounded and complete.

Proof: The largest element in the lattice is $\mathscr{X}$, while the smallest is $s(\varnothing)$. Since arbitrary intersections of closed sets are closed, condition i of Theorem 11 holds, and the lattice is complete.   ■

The first two statements in Theorem 11 differ only in that they exchange the words "meet" and "join." This is one instance of *duality* in lattices. An inspection of $L1$, $L2$, $L3$, and $L4$ shows that each statement has two parts that differ from each other only by the exchange of the symbols $\vee$ and $\wedge$. Thus any statement that is true for all lattices has a dual statement, also true for all lattices, and formed by exchanging meet and join, $\leq$ and $\geq$. An example comes from paraphrasing Theorem 11. It was proved above that the statement "If every subset of a bounded lattice has a meet, the lattice is complete" is true. Therefore the dual statement "If every subset of a bounded lattice has a join, the lattice is complete" is also true.

But more can be said. If $(L, \leq)$ is a lattice, so is $(L, \geq)$; where $x \geq y$ is defined to mean $y \leq x$. These lattices are said to be *duals* of each other.

Given a lattice L, its dual is often written as L^*, joins in L are meets in L^*, and conversely.

Definition. The lattices $(L, \vee, \wedge)$ and $(M, \vee, \wedge)$ are *isomorphic* if there is a one-one, onto mapping $f : L \to M$ such that

1. $f(x \vee y) = f(x) \vee f(y)$,
2. $f(x \wedge y) = f(x) \wedge f(y)$ for every $x, y \in L$.

As is usual, the mapping f, given above, is called an *isomorphism*, and if $L = M$, f is called an *automorphism*.

■ **LEMMA.** If f is an isomorphism from L to M, then f^{-1} is an isomorphism from M to L.

Proof: Let $a, b \in M$. Then $a = f(x)$ and $b = f(y)$ for some x, y in L, and $a \vee b = f(x) \vee f(y) = f(x \vee y)$. But $x = f^{-1}(a)$ and $y = f^{-1}(b)$. Since $a \vee b = f(x \vee y)$, we also have $x \vee y = f^{-1}(a \vee b)$. Thus $f^{-1}(a) \vee f^{-1}(b) = f^{-1}(a \vee b)$, and f^{-1} preserves joins. The proof that f^{-1} preserves meets is similar and is left as an exercise.   ■

■ **THEOREM 12.** Lattices $(L, \vee, \wedge)$ and $(M, \vee, \wedge)$ are isomorphic if and only if there is a one-one onto mapping $f : L \to M$ such that $x \leq y$ in L if and only if $f(x) \leq f(y)$ in M.

Proof: If f is an isomorphism from $(L, \vee, \wedge)$ to $(M, \vee, \wedge)$, then f is one-one and onto by definition. If $x \leq y$ in L, then $x \vee y = y$; thus $f(x) \vee f(y) = f(y)$ and $f(x) \leq f(y)$. If $f(x) \leq f(y)$ in M, then $f(x) \vee f(y) = f(y)$. By the lemma, f^{-1} preserves joins, so $x \vee y = y$ in L and $x \leq y$.

Conversely, let f be a bijection from L to M such that $x \leq y$ if and only if $f(x) \leq f(y)$ for all x, y in L. Let $s, t \in L$. Then $f(s) \leq f(s \vee t)$ and $f(t) \leq f(s \vee t)$. If $f(s), f(t) \leq a$ for some $a \in M$, then $a = f(w)$ for some w in L, and since $f(s) \leq f(w)$, $s \leq w$. Similarly, $t \leq w$, so $s \vee t \leq w$. Thus $f(s \vee t) \leq f(w) = a$, and $f(s \vee t)$ is the least upper bound of $f(s)$ and $f(t)$; therefore $f(s \vee t) = f(s) \vee f(t)$. The proof that f preserves meets is similar and is left as an exercise.   ■

That "if and only if" is needed in the last clause of Theorem 12 is evident from the following example. Let L be the lattice of subsets of a two-element set $\{a, b\}$, and let M be the lattice of the first four natural numbers $1, 2, 3, 4$, with their usual order. Define $f : L \to M$ by $f(\varnothing) = 1$, $f(\{a\}) = 2$, $f(\{b\}) = 3$, and $f(\{a, b\}) = 4$. Then, if $x \leq y$ in L, $f(x) \leq f(y)$ in M. The converse is not true, since $f(\{a\}) \leq f(\{b\})$ in M, but $\{a\} \not\subseteq \{b\}$ in L. Further, f is not an isomorphism, since $f(\{a\} \cup \{b\}) = 4$, while $f(\{a\}) \vee f(\{b\}) = 2 \vee 3 = 3$.

The examples above are illustrated in Fig. 8.1 by "Hasse diagrams." If a lattice (or poset) is small, it can sometimes be represented by making

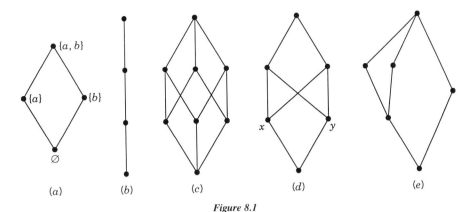

Figure 8.1

small circles or blocks for its elements, and indicating that one element is immediately below another by drawing a line upward from the smaller to the larger. The first two lattices are the ones in the example (the second is called the "four element chain"). Figure 8.1c is the lattice of subsets of a three element set (this lattice is called 2^3). The fourth poset is not a lattice because, among other things, x and y have no join.

EXERCISES

1. Show that every finite lattice is complete. Show further that if $(L, \leq)$ is a finite poset in which every pair of elements has a join, then $(L, \leq)$ is a lattice. [*Hint*: First show that every subset of L has a join; then to form $x \wedge y$, consider the join of all the elements less than or equal to both x and y.]

2. Suppose that L is a bounded lattice such that every subset of L has a join. Show that L is complete.

3. Verify that $(L, \leq)$ is a poset if and only if $(L, \geq)$ is a poset. Do the same with "lattice" replacing "poset."

4. Let $\mathcal{X}$ be a set and L the lattice of all its subsets. Show that L is isomorphic to its dual L^*, with the isomorphism given by $f(\mathcal{M}) = \mathcal{X} \setminus \mathcal{M}$.

5. Prove that if f is an isomorphism from $(L, \vee, \wedge)$ to $(M, \vee, \wedge)$, then $f^{-1}(a \wedge b) = f^{-1}(a) \wedge f^{-1}(b)$ for all $a, b \in M$.

6. Complete the proof of Theorem 12 by showing that if f is a bijection from L to M such that $x \leq y$ in L if and only if $f(x) \leq f(y)$ in M, then f preserves meets.

8.6 DISTRIBUTIVE, MODULAR, AND ATOMIC LATTICES

Much of the terminology of lattice theory comes from that of sets, and some lattice properties are suggested by the properties of the lattice $2^{\mathcal{X}}$. One such example is distributivity.

Definition. A lattice $(L, \vee, \wedge)$ is said to be *distributive* if, for all a, b, c, in L, $a \vee (b \wedge c) = (a \vee b) \wedge (a \vee c)$ and $a \wedge (b \vee c) = (a \wedge b) \vee (a \wedge c)$.

The lattices $2^{\mathscr{Q}}$ are distributive, and so is the division lattice $(\mathbb{N}, |)$. The lattice shown in Fig. 8.1e is not distributive, and neither are the lattices of flats of projective and affine space. A single example illustrates the last statement. Let ℓ, m, k, be coplanar lines in either an affine or a projective space; let $\ell \cap m \cap k = \{P\}$, and let the plane containing them be α. Then $\ell \wedge (m \vee k) = \ell \wedge \alpha = \ell$, but $(\ell \wedge m) \vee (\ell \wedge k) = \{P\} \vee \{P\} = \{P\}$.

A weakening of the distributivity condition is modularity, which will be shown to hold in the lattice of flats of any projective (but not affine) space:

Definition. A lattice $(L, \vee, \wedge)$ is said to be *modular* if, whenever a, b, and c are in L, with $a \leq c$, then $a \vee (b \wedge c) = (a \vee b) \wedge c$.

That modularity is a weakening of distributivity is shown in the following theorem:

■   **THEOREM 13.** Every distributive lattice is modular.

Proof: Suppose that L is a distributive lattice, and let a, b, and c be in L with $a \leq c$. Then, since $a \leq c$, it follows that $a \vee c = c$, and $a \vee (b \wedge c) = (a \vee b) \wedge (a \vee c) = (a \vee b) \wedge c$.   ■

The lattice (usually called M_3) shown in Fig. 8.2a is modular but not distributive; the lattice N_5 shown in Fig. 8.2b is not modular.

■   **THEOREM 14.** If $a \leq c$ in a lattice L, then for any $b \in L$, $a \vee (b \wedge c) \leq (a \vee b) \wedge c$.

Proof: For any a and b, $a \leq a \vee b$. By hypothesis, $a \leq c$; thus $a \leq (a \vee b) \wedge c$. Furthermore $(b \wedge c) \leq (a \vee b)$, and $b \wedge c \leq c$, so $(b \wedge c) \leq (a \vee b) \wedge c$. Since both a and $b \wedge c$ are beneath the element $(a \vee b) \wedge c$, so is their join, and we have $a \vee (b \wedge c) \leq (a \vee b) \wedge c$.   ■

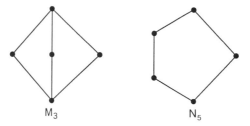

$$M_3 \qquad\qquad N_5$$

Figure 8.2

Corollary. A lattice L is modular if and only if, for all a, b, c, in L, with $a \leq c$, $(a \vee b) \wedge c \leq a \vee (b \wedge c)$.

■ **THEOREM 15.** The lattice of linear subspaces of any vector space $\mathscr{V}$ over a division ring $\mathbb{D}$ is modular.

Proof: Suppose that $\mathscr{U}$, $\mathscr{T}$, and $\mathscr{W}$ are linear subspaces of $\mathscr{V}$ with $\mathscr{U} \subseteq \mathscr{W}$. It must be shown that

$$(\mathscr{U} + \mathscr{T}) \cap \mathscr{W} \subseteq \mathscr{U} + (\mathscr{T} \cap \mathscr{W}).$$

Suppose that $\mathbf{x} \in (\mathscr{U} + \mathscr{T}) \cap \mathscr{W}$. Then $\mathbf{x} = \mathbf{u} + \mathbf{t}$, where $\mathbf{u} \in \mathscr{U}$, and $\mathbf{t} \in \mathscr{T}$, furthermore $\mathbf{x} \in \mathscr{W}$. Since $\mathbf{u} \in \mathscr{U} \subseteq \mathscr{W}$ and $\mathbf{x} \in \mathscr{W}$, then $\mathbf{x} - \mathbf{u} \in \mathscr{W}$, so $\mathbf{t} \in \mathscr{W}$. Thus $\mathbf{t} \in \mathscr{T} \cap \mathscr{W}$, and $\mathbf{x}$ is a linear combination of an element $\mathbf{u} \in \mathscr{U}$ and an element $\mathbf{t} \in \mathscr{T} \cap \mathscr{W}$. Consequently $\mathbf{x} \in \mathscr{U} + (\mathscr{T} \cap \mathscr{W})$. ■

Since the flats of any Desarguesian projective space correspond exactly to the linear subspaces of a vector space, Theorem 15 implies that the lattice of flats of any Desarguesian projective space is modular. However, a direct proof, not depending on Desargues' Theorem and therefore valid in every projective *plane* as well, is given next. The reader should note the similarity of the proofs of Theorems 15 and 16.

■ **THEOREM 16.** The lattice of flats of any projective space $(\mathscr{P}', \mathscr{L}')$ is modular.

Proof: Suppose that $\mathscr{F}$, $\mathscr{M}$, and $\mathscr{N}$ are flats with $\mathscr{F} \subseteq \mathscr{N}$. It must be shown that $(\mathscr{F} \vee \mathscr{M}) \cap \mathscr{N} \subseteq \mathscr{F} \vee (\mathscr{M} \cap \mathscr{N})$. Suppose that $\mathrm{P} \in (\mathscr{F} \vee \mathscr{M}) \cap \mathscr{N}$. Then $\mathrm{P} \in \ell(\mathrm{F}, \mathrm{M})$, where F and M are points in $\mathscr{F}$ and $\mathscr{M}$, respectively. P is also in $\mathscr{N}$, and since $\mathscr{F} \subseteq \mathscr{N}$, we have $\ell(\mathrm{P}, \mathrm{F}) \subseteq \mathscr{N}$, so $\mathrm{M} \in \mathscr{N}$. Therefore $\mathrm{P} \in \ell(\mathrm{F}, \mathrm{M})$, where $\mathrm{F} \in \mathscr{F}$ and $\mathrm{M} \in \mathscr{M} \cap \mathscr{N}$, so $\mathrm{P} \in \mathscr{F} \vee (\mathscr{M} \cap \mathscr{N})$. ■

The points in an affine or projective space play a special role in the lattice of flats of the space, as described below. As in arithmetic, the symbol $<$ means "less than but unequal to."

Definition. An element b of a lattice L is said to *cover* a (written $a \lessdot b$) if $a < b$, and there is no element c such that $a < c < b$. If L is a lattice with 0, p is an *atom* of L if $0 \lessdot p$. If L has a 1, an element h is a *coatom* if $h \lessdot 1$.

Thus an atom of a lattice is an element directly above 0, and a coatom is directly below 1. Since the points of an affine or projective space are its smallest nonempty flats, the points are the atoms of the lattice of flats in each case. The coatoms in each case are the maximal proper flats. But more can be said.

Definition. A lattice is called *atomic* if every nonzero element is the join of the atoms beneath it. (Some books use the word "atomistic" instead of "atomic.")

■ **THEOREM 17.** The lattice of flats of any projective or affine space is atomic.

Proof: If $\mathscr{F}$ is a flat of either a projective or an affine space, then $\mathscr{F}$ is the smallest flat that contains every point in $\mathscr{F}$ and is therefore the join of its set of points. ■

■ **THEOREM 18.** If L is the lattice of flats of a projective space, and p is an atom with $p \not\le a$, then $a \lessdot p \vee a$.

Proof: Clearly $a \le p \vee a$. If q is an atom under $p \vee a$ but not under a, by Chapter 7, Theorem 6, $p \vee a = q \vee a$. Thus, if $a < b < p \vee a$, choose q an atom under b but not a to get a contradiction. ■

The lattices $\mathbf{2}^{\mathscr{X}}$ are atomic, the atoms being the singleton subsets of $\mathscr{X}$. The atoms of the division lattice $(\mathbb{N}, |)$ are the prime numbers (it is notationally unfortunate that the smallest element of that lattice is the number 1; the lattice has no largest element). However, $(\mathbb{N}, |)$ is *not* atomic: The atoms under the number 12 are 2 and 3; however, $2 \vee 3 = 6 \ne 12$.

A lattice in which every pair of elements is comparable ($x \le y$ or $y \le x$) is called a *chain* (Fig. 8.1*b* shows a four-element chain). Every finite chain has exactly one atom and is not atomic. The chain of all real numbers with their usual order has no atoms; in fact it has no smallest element. The chain of all nonnegative real numbers has a smallest element, 0, but it has no atoms because there is no smallest *positive* real number. Of the lattices in Fig. 8.2., M_3 is atomic, while N_5 is not.

The properties defined so far give a partial characterization of the lattice of flats of a projective space. To complete the characterization, a lemma is needed.

■ **LEMMA.** Suppose that p, r, and s are atoms of a modular lattice L with $p \ne s$. Then, if $p \le r \vee s$, $p \vee s = r \vee s$.

Proof: Since $p \le r \vee s$, $p \vee s \le r \vee s$. If $r \wedge (p \vee s) = 0$, then $s = s \vee [r \wedge (p \vee s)] = (s \vee r) \wedge (p \vee s) \ge p$, a contradiction. Since r is an atom, $r \wedge (p \vee s) = r$, so $r \le p \vee s$. Thus $r \vee s \le p \vee s$, and $r \vee s = p \vee s$. ■

■ **THEOREM 19.** Suppose that L is an atomic modular lattice with at least three elements. Then $(\mathscr{P}', \mathscr{L}')$ satisfies the first two axioms for projective space where $\mathscr{P}'$ is the set of atoms of L and for atoms p and q, $\ell(p, q) = \{r : r$ an atom and $r \leq p \vee q\}$.

Proof: PS1. If p and q are distinct atoms of L, then both p and q are under $p \vee q$. By definition, p and q are in $\ell(p, q)$. Suppose that p and q are in $\ell(r, s)$, where r and s are distinct atoms, and assume without loss of generality that $p \neq s$. Since $p \leq r \vee s$, by the lemma $p \vee s = r \vee s$. Now $q \leq p \vee s$, and since $q \neq p$, the lemma gives $p \vee q = p \vee s = r \vee s$, and the first axiom for projective space holds.

PS2 (The Pasch Axiom). Suppose that p, q, r, s, and t are distinct atoms with $t = (p \vee q) \wedge (r \vee s)$. If $(p \vee r) \wedge (q \vee s) = 0$, we have $q = q \vee 0 = q \vee [(p \vee r) \wedge (q \vee s)] = (q \vee p \vee r) \wedge (q \vee s)$. But since $t \leq p \vee q$, $p \vee q = p \vee t$, so $(q \vee p \vee r) = p \vee t \vee r$. But $t \leq r \vee s$, so $t \vee r = t \vee s$, and $(q \vee p \vee r) = p \vee t \vee r = p \vee t \vee s$. Now $s \leq (q \vee p \vee r) \wedge (q \vee s) = q$, a contradiction. Thus $(p \vee r) \wedge (q \vee s)$ $\neq 0$, and the Pasch Axiom holds. ∎

EXERCISES

1. Show that the lattice shown in Fig. 8.1*e* is not modular.

*2. Suppose that $(L, \vee, \wedge)$ is a lattice. Prove that the following two statements are equivalent:
    a. For all $a, b, c \in L$, $a \vee (b \wedge c) = (a \vee b) \wedge (a \vee c)$.
    b. For all $a, b, c \in L$, $a \wedge (b \vee c) = (a \wedge b) \vee (a \wedge c)$.

3. Show, by producing an example, that given particular elements a, b, and c in L with $a \vee (b \wedge c) = (a \vee b) \wedge (a \vee c)$, it does *not* necessarily follow that $a \wedge (b \vee c) = (a \wedge b) \vee (a \wedge c)$.

4. Show that the lattice of flats of an affine space is never modular by considering two parallel lines ℓ and m, with a point $P \in m$.

5. Prove that the lattice of *normal* subgroups of any group is modular. Show, by finding an example, that the lattice of *all* subgroups of a group is not necessarily modular. [*Hint:* Since every subgroup of an abelian group is normal, look at nonabelian groups, such as the dihedral groups, for a counterexample.]

6. Prove that every chain is a distributive lattice.

*7. $M \subseteq L$ is a sublattice of L if, whenever a and b are in M, then $a \vee b$ and $a \wedge b$ are in M. Prove that L is modular if and only if L does not contain N_5 (Fig. 8.2*b*) as a sublattice.

*8. Write the statement of Desargues' Theorem as a lattice theoretic statement in the lattice of flats of a projective plane. Do the same for Pappus' Theorem.

8.7 COMPLETE LATTICES AND CLOSURE SPACES

It was shown in the corollary to Theorem 11 that the lattice of closed subspaces of any closure space is a complete lattice. A partial converse is proved next.

■   **THEOREM 20.** If L is a complete atomic lattice, then $L \cong \mathscr{C}(\mathscr{X}, s)$ for some closure space $(\mathscr{X}, s)$.

Proof: Let $\mathscr{X}$ be the set of atoms of L, and for $A \neq \varnothing \subseteq \mathscr{X}$, define $s(A) = \{p : p \in \mathscr{X}, p \leq \vee A\}$ and $s(\varnothing) = \varnothing$. Clearly $A \subseteq s(A)$ for every subset A of atoms, so s is increasing.

In particular, $s(A) \subseteq s(s(A))$. To show the reverse inclusion, note that $\vee\{p : p \in \mathscr{X}, p \leq \vee A\} \leq \vee A$ (since $\vee A$ is an upper bound for all the p's). Thus, if $q \leq \vee\{p : p \in \mathscr{X}, p \leq \vee A\}$, then $q \leq \vee A$. But, if $q \in s(s(A))$, then $q \in \mathscr{X}$ and $q \leq \vee s(A) = \vee\{p : p \in \mathscr{X}, p \leq \vee A\}$. Thus $q \leq \vee A$ and $q \in s(A)$. Therefore s is idempotent.

If $A \subseteq B$, then $\vee A \leq \vee B$, so $p \leq \vee A$ implies that $p \leq \vee B$. Therefore $s(A) \subseteq s(B)$, and s is isotone.

Thus s is a closure operator. To show that L is isomorphic to $\mathscr{C}(\mathscr{X}, s)$, for $A = s(A)$ define $f(A) = \vee A \in L$. If $A = s(A) \neq s(B) = B$, then without loss of generality, there is an atom $p \leq \vee A$ with $p \nleq \vee B$. Thus $\vee A \neq \vee B$, and f is one-one.

For an arbitrary element $a \in L$, let $A = \{p : p \text{ an atom and } p \leq a\}$. Since L is atomic, $\vee A = a$, so f is onto.

If $A = s(A) \subseteq s(B) = B$, then $\vee A \leq \vee B$; conversely, if lattice element a is less than or equal to b, the set of atoms under a is a subset of the set of atoms under b. Thus f and f^{-1} preserve inclusion, and f is an isomorphism.   ■

Corollary. The closure space $(\mathscr{X}, s)$ in Theorem 20 is normalized and point-closed.

Proof: Normalized is immediate from the definition, and $(\mathscr{X}, s)$ is point-closed since $s(\{p\}) = \{q \in \mathscr{X} : q \leq \vee\{p\} = p\} = \{p\}$.   ■

The counterpart to the locally finite condition for closure spaces is the property of being an algebraic lattice as defined below.

Definition. An atomic lattice L is called *algebraic* if, whenever p and $\{p_i\}_{i \in I}$ are atoms of L with $p \leq \vee_{i \in I} p_i$, there is a finite subset $F \subseteq I$ such that $p \leq \vee_{i \in F} p_i$.

■   **THEOREM 21.** Let L be a complete atomic lattice isomorphic to the point-closed closure space $\mathscr{C}(\mathscr{X}, s)$, as defined in Theorem 20. Then L is algebraic if and only if $(\mathscr{X}, s)$ is locally finite.

Proof: If $(\mathscr{X}, s)$ is locally finite and $p \leq \vee_{i \in I} p_i$, then $p \in s(\{p_i\}_{i \in I})$, so $p \in s(\{p_i\}_{i \in F})$ for some finite subset $F \subseteq I$ and L is algebraic. Conversely, if L is algebraic and $p \in s(A)$ for some subset $A \subseteq \mathscr{X}$, then $p \leq \vee_{q \in A} s(\{q\})$, so there is a finite subset of A, say, $\{q_1, \ldots, q_n\}$ with $p \leq q_1 \vee \cdots \vee q_n$. Thus $p \in s(\{q_1, \ldots, q_n\})$, and $(\mathscr{X}, s)$ is locally finite. ■

Corollary. The lattice of flats of any projective or affine space is algebraic.

The lattice corresponding to an additive closure space is a so-called biatomic lattice, defined as follows:

Definition. An atomic lattice L is *biatomic* if whenever a and b are nonzero elements of L, and p is an atom with $p \leq a \vee b$, there are atoms $a_1 \leq a$ and $b_1 \leq b$ such that $p \leq a_1 \vee b_1$.

■ **THEOREM 22.** Let L be a complete atomic lattice isomorphic to the lattice $\mathscr{C}(\mathscr{X}, s)$ as defined in Theorem 20. Then $(\mathscr{X}, s)$ is additive if and only if L is biatomic.

Proof: If $(\mathscr{X}, s)$ is additive and p is an atom of L with $p \leq a \vee b$ ($a \neq 0 \neq b$), then for A and B the sets of atoms under a and b, respectively, $p \in s(A \cup B)$. Therefore $p \in s(\{a_1, b_1\})$ for some $a_1 \in A$ and $b_1 \in B$. Thus $p \leq a_1 \vee b_1$.

Conversely, suppose that L is biatomic, and that A and B are closed subspaces of the set of atoms $\mathscr{X}$. If $p \in s(A \cup B)$, then $p \leq \vee A \vee \vee B$; thus $p \leq a_1 \vee b_1$, where $a_1 \leq \vee A$ and $b_1 \leq \vee B$. But, since A and B are closed in $\mathscr{X}$, $a_1 \in A$ and $b_1 \in B$, so $(\mathscr{X}, s)$ is additive. ■

Corollary. The lattice of flats of any projective space is biatomic. The lattice of flats of an affine space is never biatomic.

Proof: The first statement holds because projective closure spaces are additive by Theorem 4. The affine closure spaces $(\mathscr{P}, \overline{})$ are not additive: If $\ell(\mathrm{P}, \mathrm{Q})$ and m are parallel lines contained in the plane α, then $\overline{\{\mathrm{P}\} \cup m} = \alpha$, but Q is on no line through P and a point of m. ■

Every atomic modular lattice is biatomic. However, we need only a weaker statement, given in the following lemma:

■ **LEMMA.** Suppose that L is an atomic modular lattice. Then for p and q distinct atoms with $p \leq q \vee b$, there is an atom $b_1 \leq b$ such that $p \leq q \vee b_1$.

Proof: (McKenzie, McNulty, and Taylor 1987, p. 211). If $p \leq b$, take $b_1 = p$ to obtain $p \leq q \vee b_1$. Otherwise, by modularity, $[(p \vee q) \wedge b] \vee q = (p \vee q) \wedge (b \vee q) = q \vee [p \wedge (b \vee q)] = p \vee q$. Since L is atomic, there is an atom, say, b_1 under $(p \vee q) \wedge b$. If $b_1 = q$, then $q \leq b$, and $p \leq q \vee b \leq b$, a contradiction. Thus b_1 is different from p and q. But $b_1 \leq p \vee q$. Consequently by the lemma preceding Theorem 19, $p \leq q \vee b_1$. ■

■   **THEOREM 23.** Let L be an algebraic atomic modular lattice. Then $(\mathscr{P}', \mathscr{L}')$ satisfies the first two axioms for projective space where $\mathscr{P}'$ is the set of atoms of L and for atoms p and q, $\ell(p, q) = \{r : r$ an atom and $r \leq p \vee q\}$. Furthermore L is isomorphic to the lattice of flats of $(\mathscr{P}', \mathscr{L}')$.

Proof: All but the last sentence was proved in Theorem 19. Suppose that $\mathscr{F}$ is a flat of $(\mathscr{P}', \mathscr{L}')$. Since L is atomic, every element of L can be identified with the set of atoms under it. If $\mathscr{S}$ is the set of atoms under a lattice element x, and if p and q are in $\mathscr{S}$ with $r \leq p \vee q$ (r an atom), then $r \leq x$, so $\mathscr{S}$ is a flat of $(\mathscr{P}', \mathscr{L}')$. Conversely, suppose that $\mathscr{F}$ is a flat of $(\mathscr{P}', \mathscr{L}')$. We will show that $\mathscr{F}$ is exactly the set of atoms under the lattice element $x = \vee \mathscr{F}$. Clearly $\mathscr{F}$ is contained in the set of atoms under its join. Conversely, if p is an atom under $\vee \mathscr{F}$, then there are atoms $q_1, \ldots, q_n$ in $\mathscr{F}$ with $p \leq q_1 \vee \cdots \vee q_n$. If $n = 2$, $p \in \mathscr{F}$, since $\mathscr{F}$ is a flat of $(\mathscr{P}', \mathscr{L}')$. Assume that for $k \leq n - 1$, if $p \leq q_1 \vee \cdots \vee q_k$, with the $q_i \in \mathscr{F}$, then $p \in \mathscr{F}$. If $p \leq q_1 \vee \cdots \vee q_n$, then there is an atom r with $p \leq r \vee q_n$ by Lemma 3. The atom r is in $\mathscr{F}$ by the induction hypothesis, and $p \in \mathscr{F}$, since $\mathscr{F}$ is a flat.   ■

So far we have ignored axiom $PS3$, that there are at least three points on every line and at least two lines. The second part of the axiom can be stated lattice theoretically by postulating that L has atoms p, q, and r with $p \wedge (q \vee r) = 0$. The lattice theoretic equivalent to the statement that there are at least three points per line is more surprising, and it depends on the idea of the direct product of lattices.

Definition. If L and M are lattices, their *direct product* is given by $(L \times M, \leq)$, where $(a, b) \leq (c, d)$ if and only if $a \leq c$ and $b \leq d$. A lattice is called *irreducible* if it cannot be written as a direct product (except as the trivial product of itself and a one-element lattice).

It is an easy exercise to show that in $L \times M$, joins and meets are computed componentwise, so that

$$(a, b) \vee (c, d) = (a \vee c, b \vee d)$$

and

$$(a, b) \wedge (c, d) = (a \wedge c, b \wedge d).$$

If L and M are atomic lattices, the atoms of the product $L \times M$ are the elements $(p, 0)$ and $(0, q)$, where p is an atom of L and q is an atom of M. (Note that (p, q) is not an atom of $L \times M$, since $(0, 0) < (p, 0) < (p, q)$.) The join of the atoms $(p, 0)$ and $(0, q)$ is (p, q), and the element (p, q) has only the atoms $(p, 0)$ and $(0, q)$ under it. Therefore the atoms and joins of pairs of atoms of a direct product of two lattices cannot be made into a projective space because the "lines" of the form $(p, 0) \vee (0, q)$ contain only two points. From this it follows immediately that

■ **THEOREM 24.** The lattice of flats of any projective space is irreducible.

It is also true that if L is a complete, atomic, algebraic, modular lattice, then under the join of every pair of distinct atoms p and q there exists a third atom r if and only if L is irreducible. The proof is beyond our scope here, but the interested reader can consult (Birkhoff 1967, ch. IV, secs. 6, 7), or (Grätzer 1978, pp. 180–182). Thus the lattice theoretic characterization of projective geometry can be summarized in the following theorem:

■ **THEOREM 25.** L is the lattice of flats of a projective space if and only if L is complete, atomic, algebraic, modular, and irreducible.

EXERCISES

1. Suppose that L is any complete lattice. Show that L is isomorphic to $\mathscr{C}(L, s)$, where $s(A) = \{x \in L : x \leq \vee A\}$.

2. Suppose that L is an atomic modular lattice. Prove that the *atomic exchange property* holds in L, namely that for x an element of L and p, q atoms of L with $p \leq q \vee x$, but $p \not\leq x$, then $q \leq p \vee x$.

3. Suppose that L is a complete atomic lattice with $\mathscr{X}$ its set of atoms. Prove that the closure space $\mathscr{C}(\mathscr{X}, s)$ satisfies the Steinitz-Mac Lane exchange property if and only if L has the atomic exchange property.

4. Prove that in the direct product of lattices L and M, joins and meets are given as follows: $(a, b) \vee (c, d) = (a \vee c, b \vee d)$ and $(a, b) \wedge (c, d) = (a \wedge c, b \wedge d)$.

5. Prove that the product $L \times M$ of lattices L and M is modular if and only if both L and M are modular. Do the same for distributivity.

6. Prove that the direct product of nontrivial lattices L and M is atomic if and only if L and M are atomic. Is the statement true for biatomic lattices?

7. Prove that the *lexicographic product* of lattices L and M is a lattice where $(a, b) \leq (c, d)$ if $a \leq c$ or $a = c$ and $b \leq d$.

SUGGESTED READING

J. Abbott, *Sets, Lattices, and Boolean Algebras*, Boston: Allyn and Bacon, 1969.

M. K. Bennett, Affine geometry: A lattice characterization, *Proc. AMS* **88** (1983), 21–26.

G. Birkhoff, *Lattice Theory*, 3d ed., Providence: American Math Society, 1967.

B. Davey and H. Priestley, *Introduction to Lattices and Order*, Cambridge: Cambridge University Press, 1990.

G. Grätzer, *General Lattice Theory*, New York: Academic Press, 1978.

R. McKenzie, G. McNulty, and W. Taylor, *Algebras, Lattices, Varieties*, vol. 1, Monterey: Wadsworth and Brooks/Cole, 1987.

M. Stern, *Semimodular Lattices*, Stuttgart: Teubner, 1991.

CHAPTER 9

Collineations

This chapter studies the isomorphisms (called *collineations*) of affine and projective spaces, and shows that every projective isomorphism is induced by an isomorphism on the corresponding lattices of flats. It also shows (in the Fundamental Theorem of Projective Geometry) that the lattice of subspaces of any vector space determines that vector space up to isomorphism.

9.1 GENERAL COLLINEATIONS

Just as bijections which preserve addition and multiplication are the appropriate isomorphisms of rings and fields, bijections of points which preserve collinearity are the appropriate isomorphisms for both affine and projective geometries. More formally, we have the following definition:

Definition.

1. A *collineation* between the projective spaces $(\mathscr{P}', \mathscr{L}')$ and $(\mathscr{Q}', \mathscr{K}')$ is a bijection $\phi : \mathscr{P}' \to \mathscr{Q}'$ such that whenever P and Q are distinct points of $\mathscr{P}'$, then $R \in \ell(P, Q)$ if and only if $\phi(R) \in \ell(\phi(P), \phi(Q))$.

2. A *collineation* between the affine spaces $(\mathscr{P}, \mathscr{L}, \mathscr{E})$ and $(\mathscr{Q}, \mathscr{K}, \mathscr{F})$ (or the affine planes $(\mathscr{P}, \mathscr{L})$ and $(\mathscr{Q}, \mathscr{K})$) is a bijection $\phi : \mathscr{P} \to \mathscr{Q}$ such that whenever P and Q are distinct points of $\mathscr{P}$, then $R \in \ell(P, Q)$ if and only if $\phi(R) \in \ell(\phi(P), \phi(Q))$.

3. Two projective or affine spaces are said to be *isomorphic* if there is a collineation between them. This will be denoted by $(\mathscr{P}', \mathscr{L}') \cong (\mathscr{Q}', \mathscr{K}')$ or $(\mathscr{P}, \mathscr{L}, \mathscr{E}) \cong (\mathscr{Q}, \mathscr{K}, \mathscr{F})$. A collineation from a projective or affine space to itself is called an *automorphism*.

If an affine space has exactly two points on each line, then any bijection of its points is technically an automorphism, in the same way that every subset satisfies the definition for a flat. Thus, if an affine space has exactly two points on each line, we will modify the definition by assuming that every collineation maps coplanar points to coplanar points.

It is obvious from the definition that every collineation maps points to points and lines to lines. But more can be said.

■ **THEOREM 1.** Any collineation ϕ between two affine or two projective spaces is a bijection between the lines of the spaces. Conversely, if ϕ is a bijection between the points of two spaces which also defines a bijection between their lines, then ϕ is a collineation.

Proof: Let $\ell = \ell(P, Q)$ be a line in the domain space. Then for any point R in ℓ, $\phi(R) \in \ell(\phi(P), \phi(Q))$. Thus $\phi(\ell) \subseteq \ell(\phi(P), \phi(Q))$. But, if X is any point in $\ell(\phi(P), \phi(Q))$, $X = \phi(Y)$ for some Y. By definition $\phi(Y) \in \ell(\phi(P), \phi(Q))$ if and only if $X \in \ell(P, Q)$, so $\phi(\ell) = \ell(\phi(P), \phi(Q))$. If m is any line in the range space, then m is of the form $\ell(\phi(A), \phi(B))$, since ϕ is onto.

The converse follows since for any line $\ell = \ell(P, Q)$ in the domain space, $\phi(P)$ and $\phi(Q)$ are in the line $\phi(\ell)$, thus $\phi(\ell) = \ell(\phi(P), \phi(Q))$. If $R \in \ell(P, Q)$, then $\phi(R) \in \phi(\ell(P, Q)) = \ell(\phi(P), \phi(Q))$. Conversely, if $\phi(R) \in \ell(\phi(P), \phi(Q))$, then $\phi(\ell(P, Q)) = \ell(\phi(P), \phi(Q)) = \ell(\phi(R), \phi(Q)) = \phi(\ell(R, Q))$, so $\ell(P, Q) = \ell(R, Q)$ and $R \in \ell(P, Q)$. ■

■ **THEOREM 2.** Let ϕ be a collineation between the projective spaces $(\mathscr{P}', \mathscr{L}')$ and $(\mathscr{Q}', \mathscr{K}')$. Then the image of every plane in $(\mathscr{P}', \mathscr{L}')$ is a plane in $(\mathscr{Q}', \mathscr{K}')$.

Proof: If $(M, \ell(A, B))$ is a plane in $(\mathscr{P}', \mathscr{L}')$, then $(\phi(M), \ell(\phi(A), \phi(B)))$ is a plane in $(\mathscr{Q}', \mathscr{K}')$ since M, A, and B noncollinear implies that $\phi(M)$, $\phi(A)$, and $\phi(B)$ are noncollinear. If $P \in (M, \ell(A, B))$, then $P \in \ell(M, C)$ for some $C \in \ell(A, B)$. But $\phi(C) \in \ell(\phi(A), \phi(B))$. Furthermore $\phi(P) \in \ell(\phi(M), \phi(C))$, so $\phi(P) \in (\phi(M), \ell(\phi(A), \phi(B)))$. Conversely, if X is any point in $(\phi(M), \ell(\phi(A), \phi(B)))$, then $X = \phi(Q)$ for some $Q \in \mathscr{P}$. Thus $\phi(Q) \in \ell(\phi(M), \phi(D))$ for some $\phi(D) \in \ell(\phi(A), \phi(B))$. But this implies that $Q \in \ell(M, D)$ and $D \in \ell(A, B)$, so $Q \in (M, \ell(A, B))$. ■

■ **THEOREM 3.** Any collineation between the affine spaces $(\mathscr{P}, \mathscr{L}, \mathscr{E})$ and $(\mathscr{Q}, \mathscr{K}, \mathscr{F})$ defines a bijection between the planes in $\mathscr{E}$ and the planes in $\mathscr{F}$.

Proof: Because of the remarks following the definition of collineations, we can assume here that each line contains at least three points.

Suppose that α is a plane in $(\mathscr{P}, \mathscr{L}, \mathscr{E})$, and let $\ell(A, B)$ and $\ell(A, C)$ be distinct lines in α. For an arbitrary point $P \in \alpha$, if $\ell(P, B) \| \ell(A, C)$, let

D be a third point on $\ell(A, B)$. Then $\ell(P, D)$ is not parallel to $\ell(A, C)$, so $\ell(P, D) \cap \ell(A, C) = E$ for some E. Thus $P \in \ell(D, E)$, where $D \in \ell(A, B)$ and $E \in \ell(A, C)$. Therefore $\phi(P) \in \ell(\phi(D), \phi(E))$, where $\phi(D) \in \ell(\phi(A), \phi(B))$ and $\phi(E) \in \ell(\phi(A), \phi(C))$. Thus any point P in α, the plane determined by A, B, and C, is mapped to a point in the plane determined by $\phi(A)$, $\phi(B)$, and $\phi(C)$. The argument above can be reversed to show that any point in the plane $\overline{\{\phi(A), \phi(B), \phi(C)\}}$ is the image of a point in α. This is left as an exercise.   ∎

Corollary. Any collineation between affine spaces maps parallel lines to parallel lines.

Proof: By Theorem 3, any collineation maps coplanar lines to coplanar lines. If $\ell \| m$, but $\phi(\ell) \cap \phi(m) = \phi(P)$, then $P \in \ell \cap m$, a contradiction.
    ∎

Every vector space $\mathcal{V}$ over a field or a division ring can be considered as an affine space $AG(\mathcal{V})$, and its one- and two-dimensional subspaces are the points and lines of a projective space $PG(\mathcal{V})$. The linear bijections (1-1 onto linear transformations) between any two vector spaces induce both affine and projective collineations as follows:

∎ **THEOREM 4.** Let $\mathcal{V}$ and $\mathcal{W}$ be vector spaces over a division ring $\mathbb{D}$, with T a linear bijection from $\mathcal{V}$ to $\mathcal{W}$. Then T is a collineation from $AG(\mathcal{V})$ to $AG(\mathcal{W})$. Furthermore the mapping $\langle \mathbf{x} \rangle \mapsto \langle T(\mathbf{x}) \rangle$ is a collineation between the projective spaces $PG(\mathcal{V})$ and $PG(\mathcal{W})$.

Proof: T is clearly a bijection between the vectors in $\mathcal{V}$ and those in $\mathcal{W}$; namely it is a bijection between the points of $AG(\mathcal{V})$ and $AG(\mathcal{W})$. Suppose that $\mathbf{x} \in \ell(\mathbf{y}, \mathbf{z})$, where $\mathbf{x}, \mathbf{y}, \mathbf{z}, \in \mathcal{V}$. Then $\mathbf{x} = a\mathbf{y} + (1 - a)\mathbf{z}$ for some $a \in \mathbb{D}$. Since T is linear, $T(\mathbf{x}) = aT(\mathbf{y}) + (1 - a)T(\mathbf{z})$, so $T(\mathbf{x}) \in \ell(T(\mathbf{y}), T(\mathbf{z}))$. Conversely, if $T(\mathbf{w}) = aT(\mathbf{y}) + (1 - a)T(\mathbf{z}) = T(a\mathbf{y} + (1 - a)\mathbf{z})$, then $\mathbf{w} = a\mathbf{y} + (1 - a)\mathbf{z} \in \ell(\mathbf{y}, \mathbf{z})$.

For the projective collineation, the mapping $\langle \mathbf{x} \rangle \mapsto \langle T(\mathbf{x}) \rangle$ is a bijection from the points of $PG(\mathcal{V})$ to the points of $PG(\mathcal{W})$. If $\langle \mathbf{x} \rangle \in \ell(\langle \mathbf{y} \rangle, \langle \mathbf{z} \rangle)$, then $\mathbf{x}$ is in the two-dimensional subspace spanned by $\mathbf{y}$ and $\mathbf{z}$, and is therefore a linear combination $a\mathbf{y} + b\mathbf{z}$. Since T preserves linear combinations, $T(\mathbf{x})$ is in the two-dimensional subspace spanned by $T(\mathbf{y})$ and $T(\mathbf{z})$; thus $\langle T(\mathbf{x}) \rangle \in \ell(\langle T(\mathbf{y}) \rangle, \langle T(\mathbf{z}) \rangle)$. The proof of the converse is left as an exercise.   ∎

Recall from Section 8.5 that two lattices are isomorphic when there is a bijection f between them which preserves (finite) joins and meets, or equivalently such that $x \leq y$ if and only if $f(x) \leq f(y)$. We next show that the collineations between two affine or projective spaces correspond to the isomorphisms between their lattices of flats.

■   **THEOREM 5.** Suppose that ϕ is a collineation between two affine or projective spaces whose lattices of flats are L and L'. Then there is a lattice isomorphism $f: L \to L'$ such that $f_{|\text{atoms of } L} = \phi$. Conversely, if $f: L \to L'$ is a lattice isomorphism, then $f_{|\text{atoms of } L}$ is a collineation between the affine or projective spaces.

Proof: Suppose that $\mathcal{M}$ is a nonempty flat of L and define $f(\mathcal{M}) = \mathcal{N} = \{\phi(P): P \in \mathcal{M}\}$. Then, if $\phi(P)$ and $\phi(Q)$ are in $\mathcal{N}$ with $\phi(R) \in \ell(\phi(P), \phi(Q))$, we have $R \in \ell(P, Q)$, so $R \in \mathcal{M}$; thus $\phi(R) \in \mathcal{N}$, and $\mathcal{N}$ is a flat. Define $f(\varnothing) = \varnothing$. The mapping f is clearly 1-1, and it is onto since for any flat $\mathcal{N}$ in L' the set $\{P: \phi(P) \in \mathcal{N}\}$ satisfies the condition that, together with any two points, it contains every point on their line. It follows easily that $\mathcal{M} \subseteq \mathcal{M}_1$ if and only if $f(\mathcal{M}) \subseteq f(\mathcal{M}_1)$, so f is an isomorphism.

Now assume that f is a lattice isomorphism. Then p is an atom of L if and only if $f(p)$ is an atom of L', since $0 < q < p$ in L if and only if $0 < f(q) < f(p)$ in L'. Thus the restriction of f to the atoms of L is a bijection between the atoms of L and those of L'. Furthermore, for p, q, and r atoms of L, $r \leq p \vee q$ if and only if $f(r) \leq f(p) \vee f(q)$ in L'. Therefore f maps collinear points to collinear points and is a collineation.

■

This chapter will next discuss some elementary properties of collineations on projective and affine planes. Then some special collineations, namely perspectivities and projectivities of projective spaces, will be described. The Fundamental Theorem of Projective Geometry will provide the connections among vector space isomorphisms, collineations of affine and projective spaces, and lattice isomorphisms, and the chapter will conclude with a connection between Pappus' Theorem and linear transformations.

EXERCISES

1. Show that the inverse of any collineation is a collineation.

2. Complete the proofs of Theorem 3 and Theorem 4.

3. Suppose that L and L' are lattices and f is an isomorphism from L to L'. Show $f(c)$ is a coatom of L' if and only if c is a coatom of L.

4. Show that for any pair of points P and Q in the Fano plane, there is a collineation mapping P to Q. Is the collineation unique?

5. Show that the bijection $(a + bi, c + di, e + fi) \mapsto (a - bi, c - di, e - fi)$ from $\mathbb{C}^3$ to itself is *not* a linear bijection but induces a lattice automorphism on the lattice of linear subspaces of $\mathbb{C}^3$.

*6. Suppose that F is a field and σ is an automorphism of F. Show that the mapping $(x_1, \ldots, x_n) \mapsto (\sigma(x_1), \ldots, \sigma(x_n))$ induces an automorphism on the lattice of subspaces of F^n.

9.2 AUTOMORPHISMS OF PLANES

In general, the number of automorphisms of any projective plane is huge; the Fano plane has 168 automorphisms, and Marshall Hall proved that the non-Desarguesian projective Hall plane of order 9 over the Veblen-Wedderburn system $\mathcal{R}$ has 311,040 automorphisms! (Hall 1943, p. 275). See also (Stevenson 1972, sec. 12.5). It is easy to see that the composite of any two automorphisms is an automorphism, as is the inverse of any automorphism. Function composition is always associative; thus the automorphisms of any projective or affine plane form a group. There are deep results pertinent to these groups that can be found in the more specialized books (Hughes and Piper 1972) and (Stevenson 1972) but that are beyond our scope here. Instead, we will concentrate on the relationship of collineations to the construction of an affine plane from a projective one, and the relation of Desargues' Theorem to the existence of certain collineations.

Definition. A group of automorphisms of a projective or affine space is called *transitive on the points* of the space when for every pair of points P and Q there is a collineation in the group mapping P to Q. It is *transitive on the lines* of the space when for every pair of lines ℓ and m there is a collineation in the group mapping ℓ to m.

■   **THEOREM 6.** Let ℓ and m be two distinct lines of a projective plane $(\mathcal{P}', \mathcal{L}')$. Let $(\mathcal{P}, \mathcal{L})$ and $(\mathcal{Q}, \mathcal{K})$ be the affine planes obtained from $(\mathcal{P}', \mathcal{L}')$ be removing ℓ and m, respectively. Then $(\mathcal{P}, \mathcal{L}) \cong (\mathcal{Q}, \mathcal{K})$ if and only if there is an automorphism of $(\mathcal{P}', \mathcal{L}')$ that maps ℓ to m.

Proof: Suppose that ϕ is an automorphism of $(\mathcal{P}', \mathcal{L}')$ that maps ℓ to m. Then the restriction of ϕ to $\mathcal{P}$ is a bijection from $\mathcal{P}$ to $\mathcal{Q}$, and since ϕ preserves collinearity in $\mathcal{P}'$, so does its restriction.

Conversely, suppose that ψ is a collineation from $(\mathcal{P}, \mathcal{L})$ to $(\mathcal{Q}, \mathcal{K})$. Define $\phi : \mathcal{P}' \to \mathcal{Q}'$ by $\phi(A) = \psi(A)$ for $A \in \mathcal{P}$, and for $B \in \ell = \mathcal{P}' \setminus \mathcal{P}$; let Γ be the pencil of parallel lines in $(\mathcal{P}, \mathcal{L})$ whose point at infinity is B. The lines of Γ map to lines of a pencil Δ in $(\mathcal{Q}, \mathcal{K})$ by the corollary to Theorem 3. Define $\phi(B)$ as the point in m adjoined to the lines of Δ.

The mapping ϕ, restricted to $\mathcal{P}$, is a bijection from $\mathcal{P}$ to $\mathcal{Q}$. Two different points on ℓ determine different pencils in $(\mathcal{P}, \mathcal{L})$, and therefore their images are different pencils in $(\mathcal{Q}, \mathcal{K})$, corresponding to different points at infinity in m. Thus ϕ is 1-1. If C is any point in m, then C is the point at infinity for a pencil Δ of parallel lines in $(\mathcal{Q}, \mathcal{K})$. Each line of Δ is the image of a line of a pencil Γ in $(\mathcal{P}, \mathcal{L})$, and C is the image of the point at infinity on ℓ adjoined to the lines of Γ. Thus ϕ is onto.

Since ψ maps lines of $\mathscr{L}$ to lines of $\mathscr{K}$, and since for A and B in $\mathscr{P}$, with C the intersection of the projective line through A and B with the line at infinity ℓ, by definition $\phi(C)$ is the point at infinity on the line through $\phi(A)$ and $\phi(B)$. Also ϕ was defined to map points of ℓ to points of m. Thus ϕ is a bijection on the lines of $\mathscr{L}'$ and is therefore a collineation. ■

Corollary. Any two affine planes obtained by removing a line from the projective plane $(\mathscr{P}', \mathscr{L}')$ are isomorphic if and only if the group of collineations of $(\mathscr{P}', \mathscr{L}')$ is transitive on the lines of $\mathscr{L}'$.

It was stated without proof in Chapter 5 that every automorphism of the Hall plane of order 9 over the Veblen-Wedderburn system $\mathscr{R}$ maps the line $z = 0$ to itself. Therefore at least two nonisomorphic affine planes can be constructed from that projective plane. In fact, the determination of the collineation groups of the four non-isomorphic projective planes of order 9 leads to the conclusion that there are exactly seven non-isomorphic affine planes of order 9. The Hall plane is non-Desarguesian; it will be shown below that the group of automorphisms of any Desarguesian projective plane is transitive on the points and on the lines.

Definition. A *central collineation* of a projective plane $(\mathscr{P}', \mathscr{L}')$ is an automorphism that has the following properties:

CC1. There is a point C which is fixed by ϕ (i.e., $\phi(C) = C$).
CC2. For every point A $\neq$ C, $\phi(A) \in \ell(C, A)$. The point C is called the *center* of ϕ.

The identity is always a central collineation. In general, however, the center is not the only point fixed by a central collineation ϕ. It will be proved in Theorem 8 that if ϕ is not the identity, there is a unique line ℓ, called the *axis* of ϕ, such that $\phi(A) = A$ for all points A in ℓ.

Definition. A line ℓ such that $\phi(\ell) = \ell$, is called a *fixed line* for an automorphism ϕ. If, in addition, $\phi(A) = A$ for all A $\in \ell$, then ℓ is said to be *fixed pointwise* by ϕ.

■ **THEOREM 7.** If an automorphism of a projective plane fixes two lines pointwise, it is the identity.

Proof: Suppose that ℓ and m are fixed pointwise by ϕ. Let P be a point on neither ℓ nor m. Let A and B be distinct points of ℓ, and let $\ell(P, A) \cap m = A'$ and $\ell(P, B) \cap m = B'$, as shown in Fig. 9.1. Then

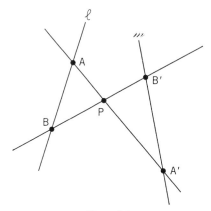

Figure 9.1

$\phi(\ell(A, A')) = \ell(\phi(A), \phi(A')) = \ell(A, A')$, and similarly $\phi(\ell(B, B')) = \ell(B, B')$. But $P = \ell(A, A') \cap \ell(B, B')$, so $\phi(P) \in \ell(\phi(A), \phi(A')) \cap \phi(\ell(B, B')) = \ell(A, A') \cap \ell(B, B') = P$, and ϕ is the identity. ∎

Corollary. Let ϕ and ψ be collineations from the projective plane $(\mathscr{P}', \mathscr{L}')$ to $(\mathscr{C}', \mathscr{K}')$. If, for distinct lines ℓ and m in $\mathscr{L}'$, $\phi_{|\ell} = \psi_{|\ell}$, and $\phi_{|m} = \psi_{|m}$, then $\phi = \psi$.

Proof: The composite $\psi^{-1} \circ \phi : \mathscr{P}' \to \mathscr{P}'$ fixes ℓ and m pointwise, and therefore it is the identity. ∎

■ **THEOREM 8.** Let ϕ be a central collineation of the projective plane $(\mathscr{P}', \mathscr{L}')$ with center C, which is not the identity. Then there is a unique line ℓ that is fixed pointwise by ϕ.

Proof: If a line ℓ, not containing C, is fixed by ϕ, then for $A \in \ell$, $\phi(A) \in \ell(C, A) \cap \ell = A$, so ℓ is fixed pointwise. Therefore at most one line not containing C is fixed by ϕ. Suppose that m is a line not containing C and not fixed by ϕ. Let $X = m \cap \phi(m)$, as shown in Fig. 9.2. Then $\phi(X) \in \ell(C, X) \cap \phi(m) = X$. Let Y be a third point on $\ell(X, C)$, let k be a line through Y not fixed by ϕ, and let $Z = k \cap \phi(k)$. Then Z is fixed by ϕ. Thus $\ell(X, Z)$ is fixed by ϕ, and it is fixed pointwise if $C \notin \ell(X, Z)$.

If $Z \in \ell(X, C)$, then $Z = k \cap \ell(X, C) = Y$. If $\{X, Z, C\} = \ell(X, Z)$, then $\ell(X, Z)$ is fixed pointwise. Otherwise, let Y' be a fourth point on $\ell(X, Z)$, and construct k' and Z' as above. If $Z' \notin \ell(X, C)$, then $\ell(X, Z')$ and $\ell(X, Y')$ are pointwise fixed, a contradiction. Therefore $Z' \in \ell(X, C)$, $Z' = Y'$, and every point of $\ell(X, C)$ is fixed by ϕ. ∎

Finally, the existence of central collineations of projective planes is connected to Desargues' Theorem as follows:

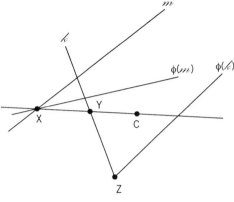

Figure 9.2

■ **THEOREM 9.** For every point C and line ℓ of a Desarguesian projective plane $(\mathscr{P}', \mathscr{L}')$ that is different from the Fano plane, there is a central collineation that is different from the identity whose center is C and whose axis is ℓ.

Proof: Let $X \neq C$, and let X' be a third point on $\ell(X, C)$ but not on ℓ. Define $\phi(C) = C$, $\phi(M) = M$ for all M in ℓ, and $\phi(X) = X'$. If $Y \notin \ell(X, C)$, define $\ell(X, Y) \cap \ell = M$, and define $\phi(Y) = Y' = \ell(Y, C) \cap \ell(X', M)$. In particular, note that $\ell(X, Y) \cap \ell(X', Y') \in \ell$; thus $\phi(Y)$ is the unique point Y' such that $Y' \in \ell(X, C)$ and $\ell(X, Y) \cap \ell(X', Y') \in \ell$.

If no three of X, Y, Z, and C are collinear, with $\ell(X, Y) \cap \ell = M$ and $\ell(X, Z) \cap \ell = N$, then $\phi(Z) = Z' = \ell(C, Z) \cap \ell(X', N)$, as shown in Fig. 9.3 both for $C \in \ell$ and for $C \notin \ell$. But by Desargues' Theorem $\ell(Y, Z) \cap \ell(Y', Z')$ is collinear with M and N, and it is therefore contained in the axis ℓ. Thus $\phi(Z)$ can be defined as the unique point Z' in $\ell(C, Z)$ such that either $\ell(X, Z) \cap \ell(X', Z') \in \ell$, or $\ell(Y, Z) \cap \ell(Y', Z') \in \ell$.

In particular, if $P \in \ell(C, X)$, take Y not on $\ell(C, X)$, define Y' as above, and define $\phi(P)$ as the unique point P' such that $P' \in \ell(C, P)$ and $\ell(P, Y) \cap \ell(P', Y') \in \ell$.

The proof that ϕ is one-one and onto is left as an exercise.

To show that ϕ is a collineation, let P, Q, and R be collinear points on line m, and let $m \cap \ell = H$. By the remarks above, given P' and Q', $\phi(R)$ can be defined as $\ell(H, P') \cap \ell(C, R)$ or as $\ell(H, Q') \cap \ell(C, R)$; therefore R', P', and Q' are collinear, and ϕ is a collineation.   ■

Corollary. The group of automorphisms of any Desarguesian projection plane is transitive on the points and transitive on the lines.

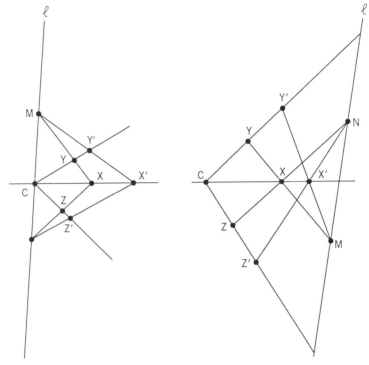

Figure 9.3

Proof: If the plane is not the Fano plane, suppose that X and X′ are distinct points. Let C be a third point on $\ell(X, X')$. By the proof of Theorem 9, there is a central collineation, with center C mapping X to X′.

Suppose that m and k are distinct lines intersecting at P. Let ℓ be a third line through P. Let X and Y be distinct points of m, let X′ and Y′ be

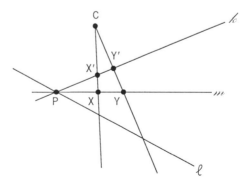

Figure 9.4

distinct points of ℓ, and let $\ell(X, X') \cap \ell(Y, Y') = C$. The central collineation with center C and axis ℓ maps X to X' and Y to Y'; thus it maps $m = \ell(X, Y)$ to $k = \ell(X', Y')$, as shown in Fig. 9.4.

The case where the plane is the Fano plane is left as an exercise. ■

EXERCISES

*1. Show that for each pair of noncollinear triples P_1, P_2, P_3 and Q_1, Q_2, Q_3 in the Fano plane, there is one and only one automorphism that maps P_1 to Q_1, P_2 to Q_2, and P_3 to Q_3. Infer that the group of automorphisms of the Fano plane is transitive on its points and its lines.

*2. Use the result in Exercise 1 to show that there are 168 automorphisms of the Fano plane. [*Hint*: Compute the number of noncollinear triples.]

3. a. Show that if $C \notin \ell$ in the Fano plane, the only central collineation with C as center and ℓ as axis is the identity.

   b. Show that if $C \in \ell$ in the Fano plane, there is exactly one central collineation other than the identity with C as center and ℓ as axis.

4. Show by using Theorem 9 that the group of automorphisms of the real projective plane is transitive on the points and lines.

5. Show by using Theorem 9 that the automorphism group of the projective plane of order 3 is transitive on the points and lines.

6. The points and lines of the projective plane $(\mathscr{P}', \mathscr{L}')$ of order 4 are listed below:

$$\mathscr{P}' = \{A, B, C, D, E, F, G, H, I, J, K, L, M, N, O, P, Q, R, S, T, U\},$$

$\mathscr{L}'$ consists of the lines

{OIDQF}	{ECPQA}	{QKMJB}	{GHQLN}	{OAJHT}	{TGIPB}
{FTCML}	{EDKNT}	{QRSTU}	{HEIMU}	{UFBAN}	{CGKOU}
{DJLPU}	{RHBCD}	{EFRGJ}	{IARKL}	{MNOPR}	{HFKPS}
{BEOSL}	{JNCIS}	{AGDMS}			

   a. Suppose that ϕ is a central collineation of $(\mathscr{P}', \mathscr{L}')$ with center C and axis {EDKNT}. Suppose further that $\phi(A) = P$. Compute $\phi(M)$.

   b. Suppose that ψ is a central collineation of $(\mathscr{P}', \mathscr{L}')$ with center C and axis {CGKOU}. Suppose further that $\psi(A) = P$. Compute $\psi(M)$.

7. Complete the proof of Theorem 9 by showing that the central collineation ϕ defined there is one-one and onto.

9.3 PERSPECTIVITIES OF PROJECTIVE SPACES

Central collineations of projective planes are examples of a more general type of automorphism on projective spaces, namely the perspectivities. Here maximal proper flats will replace lines as "axes"; thus we prove what

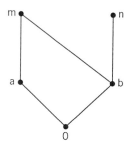

Figure 9.5

was stated in Chapter 5, that every projective space contains maximal proper flats. This is done using Zorn's Lemma stated below.

Recall from Section 8.6 that a *chain* is a lattice in which every pair of elements is comparable. A Hasse diagram of the four-element chain is given in Fig. 8.1*b*. A *maximal element* in a poset $(P, \leq)$ is an element m such that there is no $x > m$. Note that a maximal element need not be a largest element—Fig. 9.5 shows a very simple poset that has two maximal elements, m and n. The sets $\{0, m\}$, $\{0, b\}$, $\{0, a, m\}$, $\{0, b, m\}$, and $\{0, b, n\}$ are some of the chains in P.

■   **ZORN'S LEMMA.** Suppose that $(P, \leq)$ is a nonempty poset. If every chain in P has an upper bound in P, then P has a maximal element.

Zorn's Lemma is one of the equivalents to the Axiom of Choice, and as such it is assumed rather than proved.

■   **THEOREM 10.** Every projective space has a maximal proper flat.

Proof: Let X be a point in a projective space, and let $(P, \subseteq)$ be the poset of flats that do not contain X. Since there is at least one line not containing X, P is not empty. Suppose that $\mathscr{C}$ is a chain in P. We claim that $\mathscr{N} = \{P : P \in \mathscr{F} \in \mathscr{C}\}$ is a flat. For P, Q in $\mathscr{N}$ and $R \in \ell(P, Q)$, $P \in \mathscr{F}_1$ and $Q \in \mathscr{F}_2$, with $\mathscr{F}_1$ and $\mathscr{F}_2$ in $\mathscr{C}$. Then either $\mathscr{F}_1 \subseteq \mathscr{F}_2$ or $\mathscr{F}_2 \subseteq \mathscr{F}_1$, since $\mathscr{C}$ is a chain. Now P and Q are both in $\mathscr{F}_i$ ($i = 1$ or 2), so $R \in \mathscr{F}_i$ and $\mathscr{N}$ is a flat. It must be shown that $\mathscr{N} \in P$. But this follows easily for, if X were in $\mathscr{N}$, then X would be in some flat $\mathscr{F}$ contained in $\mathscr{C}$, a contradiction. Thus P has a maximal element, $\mathscr{M}$.

It must be shown that $\mathscr{M}$, which is maximal in P, is a maximal proper flat of $(\mathscr{P}', \mathscr{L}')$. This is done by showing that $\overline{\{X\} \cup \mathscr{M}} = \mathscr{P}'$ is the only flat containing $\mathscr{M}$. Suppose that $Y \in \mathscr{P}' \backslash \mathscr{M}$. Then $\overline{\{Y\} \cup \mathscr{M}}$ contains $\mathscr{M}$ as a proper subset, and by the maximality of $\mathscr{M}$, $X \in \overline{\{Y\} \cup \mathscr{M}}$. Therefore $Y \in \overline{\{X\} \cup \mathscr{M}}$, and $\overline{\{Y\} \cup \mathscr{M}} = \overline{\{X\} \cup \mathscr{M}}$. Further, since Y was an arbitrary point of $\mathscr{P}' \backslash \mathscr{M}$, it follows that $\overline{\{X\} \cup \mathscr{M}} = \mathscr{P}'$ is the only flat properly containing $\mathscr{M}$. Consequently $\mathscr{M}$ is a maximal proper flat of $(\mathscr{P}', \mathscr{L}')$.   ■

Definition. A maximal proper flat of a projective space is also called a *hyperplane*.

In Theorem 10 more was actually proved than was claimed, as the following corollary demonstrates:

Corollary. For every point X in the projective space $(\mathscr{P}', \mathscr{L}')$, there is a hyperplane that does not contain X.

Every affine space also has maximal proper flats; the proof, which also depends on Zorn's Lemma, is somewhat more complicated and will not be presented here.

Recall that a coatom of a bounded lattice L is an element h covered by 1 (symbolically $h \lessdot 1$). It is also an atom in the dual lattice L^*, and the coatoms of the lattice of flats of an affine or a projective space are its maximal proper flats, or hyperplanes.

An analogue to Theorem 7 can be proved using hyperplanes.

■   **THEOREM 11.** If an automorphism of a projective space fixes a hyperplane pointwise and fixes two additional points, then it is the identity.

Proof: Suppose that ϕ is an automorphism of $(\mathscr{P}', \mathscr{L}')$ that fixes the hyperplane $\mathscr{H}$ pointwise, and suppose further that A and B are fixed points not in $\mathscr{H}$. Let X be any point not in $\ell(A, B)$. Then, by Theorem 13 of Chapter 7, $\ell(A, X) \cap \mathscr{H} = Y$ for some point Y, and $\ell(B, X) \cap \mathscr{H} = Z$. Then $\phi(\ell(A, X)) = \phi(\ell(X, Y)) = \ell(X, Y) = \ell(A, X)$. Similarly ϕ fixes $\ell(B, X)$. Then $\phi(X) \in \ell(A, X) \cap \ell(B, X) = X$. If X is on $\ell(A, B)$, take B' not on $\ell(A, B)$. Then B' is fixed by ϕ, $X \notin \ell(A, B')$, so the argument above shows that X is fixed by ϕ.   ■

Corollary. If two automorphisms of a projective space agree on a hyperplane and two additional points, they are identical.

Definition. A *perspectivity* on a projective space is an automorphism ϕ, which has a fixed point C (called its *center*) and fixes every line through C. (Thus a central collineation of a projective plane is a perspectivity). A *projectivity* is a composite of a finite number of perspectivities.

The rest of this section will be devoted to proving an analogue to Theorem 8 for projective spaces. First some lemmas are needed.

■   **LEMMA 1.** Suppose that L and L' are lattices and that $f: L \to L'$ is an isomorphism. Then h is a coatom of L if and only if $f(h)$ is a coatom of L'.

Proof: Suppose that h is a coatom of L, and suppose that $f(h) < d < 1$ in L'. Since f is onto, $d = f(e)$ for some $e \in L$; also $1 = f(1)$. Thus $f(h) < f(e) < f(1)$, so in L, $h < e < 1$, a contradiction. The converse holds since the inverse of f is a lattice isomorphism from L' to L.   ∎

Corollary. Any collineation between projective or affine spaces maps hyperplanes to hyperplanes.

■   **LEMMA 2.** Suppose that $\mathscr{H}$ and $\mathscr{K}$ are two distinct hyperplanes in the nonplanar projective space $(\mathscr{P}', \mathscr{L}')$, and let C be a point not in their intersection. Then there are at least two lines through C that do not intersect $\mathscr{H} \cap \mathscr{K}$.

Proof: If $\mathscr{H} \neq \mathscr{K}$, then their intersection, which we will call $\mathscr{N}$, is strictly contained in each of them and therefore is not a hyperplane. If every line through C intersects $\mathscr{N}$, let X be an element of $\mathscr{P}' \setminus \mathscr{N}$. Then $\ell(X, C) \cap \mathscr{N} = Y$ for some Y, and by the Steinitz-Mac Lane exchange property (Theorem 6 of Chapter 8), $X \in \ell(C, Y) \subseteq \overline{\{C\} \cup \mathscr{N}}$, thus $\mathscr{P}' = \overline{\{C\} \cup \mathscr{N}}$. But in the lattice of flats of $(\mathscr{P}', \mathscr{L}')$, $\overline{\{C\} \cup \mathscr{N}}$ covers $\mathscr{N}$; therefore $\mathscr{N}$ is a hyperplane, a contradiction.

If there is a unique line $\ell(A, C)$ through C whose intersection with $\mathscr{N}$ is empty, let $A \in \ell(B, D) \neq \ell(A, C)$. Then $\ell(C, B) \cap \mathscr{N} = X$ and $\ell(C, D) \cap \mathscr{N} = Y$ for some X and Y. In the plane $\overline{\{C, B, D\}}$, the coplanar lines $\ell(A, C)$ and $\ell(X, Y)$ intersect in some Z in $\mathscr{N}$, a contradiction.   ∎

■   **LEMMA 3.** If $\mathscr{H}$ is a hyperplane of the projective space $(\mathscr{P}', \mathscr{L}')$, and α is a plane not contained in $\mathscr{H}$, then α intersects $\mathscr{H}$ in a line.

Proof: First, α intersects $\mathscr{H}$ in at most one line. For, if there were three noncollinear points in $\alpha \cap \mathscr{H}$, they would determine α; thus $\alpha \subseteq \mathscr{H}$, a contradiction. Assume that $\alpha = \overline{\{A, B, C\}}$. If either $\ell(A, B)$ or $\ell(A, C)$ is in $\mathscr{H}$, we are done. Otherwise, $\ell(A, B) \cap \mathscr{H} = D$ and $\ell(A, C) \cap \mathscr{H} = E$ for some points D and E by Theorem 12 of Chapter 7; thus $\ell(D, E) \subseteq \alpha \cap \mathscr{H}$, and by the remarks above, $\ell(D, E) = \alpha \cap \mathscr{H}$.   ∎

■   **THEOREM 12.** If ϕ is a perspectivity of the projective space $(\mathscr{P}', \mathscr{L}')$, with center C, then there is a unique hyperplane $\mathscr{H}$ that is fixed pointwise by ϕ.

Proof: The restriction of ϕ to any plane containing C is a central collineation of that plane, with center C. Suppose that $\mathscr{H}$ is a hyperplane not containing C but fixed by ϕ. Let P and Q be any two points of $\mathscr{H}$. Then in the plane $\overline{\{C, P, Q\}}$ the mapping ϕ has a unique axis ℓ, which is fixed pointwise by ϕ. If $\ell \neq \ell(P, Q)$, then ϕ fixes $\mathscr{H}$ and two points of ℓ pointwise, and ϕ is the identity, a contradiction. Thus $\ell = \ell(P, Q)$, and P and Q are fixed by ϕ. Since P and Q were arbitrary points in $\mathscr{H}$, we have shown that any hyperplane not containing C that is fixed by ϕ is fixed pointwise by ϕ. Therefore there is at most one hyperplane not containing C fixed by ϕ.

Let $\mathcal{H}$ be a hyperplane that does not contain C and is not fixed by ϕ. Let $\mathcal{H} \cap \phi(\mathcal{H}) = \mathcal{N}$. If $A \in \mathcal{N} \subseteq \mathcal{H}$, then $\phi(A) \in \phi(\mathcal{H})$; also $\phi(A) \in \ell(A, C)$. But $\ell(A, C) \cap \phi(\mathcal{H}) = A$, therefore $\phi(A) = A$ and $\mathcal{N}$ is fixed pointwise by ϕ.

By Lemma 2, there are at least two lines, ℓ and m, through C that do not intersect $\mathcal{N}$. The restriction of ϕ to the plane determined by the intersecting lines ℓ and m has an axis k. Let $\mathcal{K} = \mathcal{N} \vee k$, and let $X \in \mathcal{K}$. Then $X \in \ell(Y, Z)$, where $Y \in \mathcal{N}$ and $Z \in k$. Suppose that C is not on $\ell(Y, Z)$. Then $\phi(X) \in \ell(X, C) \cap \ell(\phi(Y), \phi(Z)) = \ell(X, C) \cap \ell(Y, Z) = X$, and X is fixed by ϕ. If $C \in \ell(Y, Z)$, take a different line through X in $\mathcal{N} \vee k$. Each point of that line, other than X, is fixed by ϕ; thus X is fixed by ϕ. Therefore $\mathcal{N} \vee k$ is fixed pointwise by ϕ. It is left as an exercise to show that $\mathcal{N} \vee k$ is a hyperplane. ∎

EXERCISES

1. Use Zorn's Lemma to prove that every vector space has a basis.

2. Use Zorn's Lemma to show that each point of any projective space is contained in a hyperplane.

3. Suppose that $\mathcal{H}$ and $\mathcal{K}$ are hyperplanes of a projective space $(\mathcal{P}', \mathcal{L}')$. Let $\mathcal{N} = \mathcal{H} \cap \mathcal{K}$, and let X be a point not contained in $\mathcal{N}$. Show that $\mathcal{N} \lessdot \mathcal{N} \vee \{X\} \lessdot \mathcal{P}'$ in the lattice of flats of $(\mathcal{P}', \mathcal{L}')$.

4. Show that the flat $\mathcal{N} \vee k$ of Theorem 12 is a hyperplane.

5. Show that the image under any collineation of a Desarguesian projective or affine plane is also Desarguesian.

6. Show that if an affine space satisfies Pappus' Theorem, so does its image under any collineation.

7. Given a point C and a hyperplane $\mathcal{H}$ in the Desarguesian projective space $(\mathcal{P}', \mathcal{L}')$, with at least four points on each line, show that there is a perspectivity with center C and fixed hyperplane $\mathcal{H}$. [*Hint*: See the proof of Theorem 9.]

*8. Show that every affine space has a maximal proper flat. [*Hint*: Enlarge the space to a projective space, and use Theorem 10.]

9.4 THE FUNDAMENTAL THEOREM
OF PROJECTIVE GEOMETRY

It was shown in Section 9.1 that every linear bijection between the vector spaces $\mathcal{V}$ and $\mathcal{W}$ induces a collineation between the affine spaces $AG(\mathcal{V})$ and $AG(\mathcal{W})$, as well as between the projective spaces $PG(\mathcal{V})$ and $PG(\mathcal{W})$. In this section it will be shown that a larger class, the *semilinear bijections*, induces collineations, and conversely that every collineation is induced by such a bijection.

Definition. Let $\mathscr{V}$ and $\mathscr{V}'$ be vector spaces over the division rings $\mathbb{D}$ and $\mathbb{D}'$. A *semilinear bijection* from $\mathscr{V}$ to $\mathscr{V}'$ is a pair (S, σ), where S is a bijection from $\mathscr{V}$ to $\mathscr{V}'$ and σ is an isomorphism from $\mathbb{D}$ to $\mathbb{D}'$, satisfying the following conditions:

*SB*1. $S(\mathbf{x} + \mathbf{y}) = S(\mathbf{x}) + S(\mathbf{y})$ for all $\mathbf{x}, \mathbf{y} \in \mathscr{V}$.

*SB*2. $S(a\mathbf{x}) = \sigma(a)S(\mathbf{x})$ for all $a \in \mathbb{D}$ and $\mathbf{x} \in \mathscr{V}$.

If S is a semilinear bijection, so is (S^{-1}, σ^{-1}) because S^{-1} preserves addition of vectors, and $S^{-1}(a\mathbf{x}) = \sigma^{-1}(a)S^{-1}(\mathbf{x})$. For example, let $\mathscr{V} = \mathbb{D}^n$, and let σ be any automorphism of $\mathbb{D}$ (e.g., complex conjugation when $\mathbb{D} = \mathbb{C}$). Define $S(x_1, \ldots, x_n) = (\sigma(x_1), \ldots, \sigma(x_n))$. Since σ preserves addition of the scalars x_i, S preserves vector addition. Furthermore $S(a\mathbf{x}) = S(a(x_1, \ldots, x_n)) = S(ax_1, \ldots, ax_n) = (\sigma(ax_1), \ldots, \sigma(ax_n)) = (\sigma(a)\sigma(x_1), \ldots, \sigma(a)\sigma(x_n)) = \sigma(a)(\sigma(x_1), \ldots, \sigma(x_n)) = \sigma(a)S(\mathbf{x})$.

■ **THEOREM 13.** Every semilinear bijection (S, σ) from $\mathscr{V}$ to $\mathscr{V}'$ induces a collineation from $AG(\mathscr{V})$ to $AG(\mathscr{V}')$ and one from $PG(\mathscr{V})$ to $PG(\mathscr{V}')$.

Proof: By definition, S is a bijection from the points of $AG(\mathscr{V})$ to those of $AG(\mathscr{V}')$. If $\mathbf{x}$, $\mathbf{y}$, and $\mathbf{z}$ are collinear points in $AG(\mathscr{V})$ (i.e., are vectors in $\mathscr{V}$ such that $\mathbf{z} = \mathbf{y} + t(\mathbf{y} - \mathbf{x})$), then $S(\mathbf{z}) = S(\mathbf{y}) + \sigma(t)(S(\mathbf{y}) - S(\mathbf{x}))$. Therefore $S(\mathbf{x})$, $S(\mathbf{y})$, and $S(\mathbf{z})$ are collinear in $AG(\mathscr{V}')$ and S induces a collineation from $AG(\mathscr{V})$ to $AG(\mathscr{V}')$.

The mapping $\langle \mathbf{x} \rangle \mapsto \langle S(\mathbf{x}) \rangle$ is well defined since if $\langle \mathbf{x} \rangle = \langle a\mathbf{x} \rangle$, $S(a\mathbf{x}) = \sigma(a)S(\mathbf{x})$ so that $\langle S(\mathbf{x}) \rangle = \langle S(a\mathbf{x}) \rangle$. It is a bijection from the one-dimensional subspaces of $\mathscr{V}$ to those of $\mathscr{V}'$, since S is a bijection on the vectors. If $\langle \mathbf{x} \rangle$, $\langle \mathbf{y} \rangle$, and $\langle \mathbf{z} \rangle$ are collinear in $PG(\mathscr{V})$, then $\mathbf{z} = a\mathbf{x} + b\mathbf{y}$ for some a and b in $\mathbb{D}$. Consequently $S(\mathbf{z}) = \sigma(a)S(\mathbf{x}) + \sigma(b)S(\mathbf{y})$, therefore $\langle S(\mathbf{x}) \rangle$, $\langle S(\mathbf{y}) \rangle$, and $\langle S(\mathbf{z}) \rangle$ are collinear in $PG(\mathscr{V}')$. The converse holds since (S^{-1}, σ^{-1}) is a semilinear bijection. ■

Corollary. Every semilinear bijection (S, σ) from $\mathscr{V}$ to $\mathscr{V}'$ induces a lattice isomorphism from the lattice of flats of $AG(\mathscr{V})$ to the lattice of flats of $AG(\mathscr{V}')$ and a lattice isomorphism from the lattice of flats of $PG(\mathscr{V})$ to the lattice of flats of $PG(\mathscr{V}')$.

The hyperplanes in any projective space $(\mathscr{P}', \mathscr{L}')$ are the coatoms (elements covered by the largest element 1) in its lattice of flats. The lattice of flats of an affine space can be recovered from the lattice of flats of a projective space by removing a coatom and all the nonzero lattice elements under it. More formally,

■ **THEOREM 14.** Suppose that $(L, \vee_L, \wedge_L)$ is the lattice of flats of a projective space and h is a coatom of L. Then $(M, \vee_M, \wedge_M)$ is the lattice of flats of an affine space, where

$$M = L \setminus \{x : x \neq 0 \text{ and } x \leq h\} = \{x \in L : x = 0 \text{ or } x \not\leq h\},$$

$$a \vee_M b = a \vee_L b,$$

$$a \wedge_M b = \begin{cases} a \wedge_L b & \text{if } a \wedge_L b \not\leq h, \\ 0 & \text{if } a \wedge_L b \leq h. \end{cases}$$

Proof: The coatom h corresponds to a hyperplane $\mathscr{H}$ in $(\mathscr{P}', \mathscr{L}')$. Let $(\mathscr{P}, \mathscr{L}, \mathscr{E})$ be the affine space obtained by removing $\mathscr{H}$ from $(\mathscr{P}', \mathscr{L}')$. If $\mathscr{F}$ is a nonempty flat of $(\mathscr{P}, \mathscr{L}, \mathscr{E})$, then there is a unique flat $\mathscr{F}'$ of $(\mathscr{P}', \mathscr{L}')$ such that $\mathscr{F} = \mathscr{F}' \setminus (\mathscr{F}' \cap \mathscr{H})$, or more simply $\mathscr{F}' \setminus \mathscr{H}$. Conversely, every projective flat $\mathscr{G}'$ not contained in $\mathscr{H}$ corresponds to the unique flat $\mathscr{G}' \setminus \mathscr{H}$. Thus the flats of $(\mathscr{P}, \mathscr{L}, \mathscr{E})$ are in one-one correspondence with the elements of M, the empty flat corresponding to 0.

If $\mathscr{F}'$ and $\mathscr{G}'$ are projective flats not contained in $\mathscr{H}$, we claim that $(\mathscr{F}' \vee_L \mathscr{G}') \setminus \mathscr{H} = \mathscr{F}' \setminus \mathscr{H} \vee_M \mathscr{G}' \setminus \mathscr{H}$. If $P \in (\mathscr{F}' \vee_L \mathscr{G}') \setminus \mathscr{H}$, then $P \in \ell'(F, G)$ where $F \in \mathscr{F}'$ and $G \in \mathscr{G}'$. If neither F nor G is in $\mathscr{H}$, then $F \in \mathscr{F}' \setminus \mathscr{H}$, $G \in \mathscr{G}' \setminus \mathscr{H}$ so that $P \in \mathscr{F}' \setminus \mathscr{H} \vee_M \mathscr{G}' \setminus \mathscr{H}$. At most one of F and G is in $\mathscr{H}$, since $P \notin \mathscr{H}$. If $G \in \mathscr{H}$, take G_1 to be a point of $\mathscr{G}'$ not in $\mathscr{H}$, and let m' be the projective line $\ell'(G, G_1)$. Then the affine line $m = m' \setminus \{G\}$ is parallel to $\ell(P, F)$, so m, P and F are coplanar. Thus P is in the affine span of F and m, namely $\{F\} \vee_M m$, where $F \in \mathscr{F}' \setminus \mathscr{H}$ and $m \subseteq \mathscr{G}' \setminus \mathscr{H}$. Conversely, $\mathscr{F}' \setminus \mathscr{H}$ and $\mathscr{G}' \setminus \mathscr{H}$ are both subsets of in $(\mathscr{F}' \vee_L \mathscr{G}') \setminus \mathscr{H}$, so their affine join is a subset of $(\mathscr{F}' \vee_L \mathscr{G}') \setminus \mathscr{H}$ as well.

If $\mathscr{F} = \mathscr{F}' \setminus \mathscr{H}$ and $\mathscr{G} = \mathscr{G}' \setminus \mathscr{H}$ are flats of $(\mathscr{P}, \mathscr{L}, \mathscr{E})$, with $\mathscr{F}' \cap \mathscr{G}'$ not contained in $\mathscr{H}$, then $\mathscr{F}' \setminus \mathscr{H} \cap \mathscr{G}' \setminus \mathscr{H} = (\mathscr{F}' \cap \mathscr{G}') \setminus \mathscr{H}$; thus $\mathscr{F}' \setminus \mathscr{H} \wedge_M \mathscr{G}' \setminus \mathscr{H} = (\mathscr{F}' \wedge_L \mathscr{G}') \setminus \mathscr{H}$. If $\mathscr{F}' \cap \mathscr{G}' \subseteq \mathscr{H}$, then $\mathscr{F}' \setminus \mathscr{H}$ and $\mathscr{G}' \setminus \mathscr{H}$ have no affine points in common, and their meet in M is 0. ■

We can now prove that any collineation between Desarguesian affine spaces is induced by a semilinear bijection.

Recall that in order to coordinatize a Desarguesian affine space $(\mathscr{P}, \mathscr{L}, \mathscr{E})$, we selected a point O, and on each line ℓ through O, we chose a multiplicative identity point I_ℓ. We saw in Theorem 11 of Chapter 4 that if $\ell = \ell(O, I_\ell)$ and $m = \ell(O, I_m)$, then ℓ and m are isomorphic as division rings. The isomorphism, denoted $\phi_{\ell m}$, maps I_ℓ to I_m, and any point A different from O and I_ℓ to the point A' in m such that

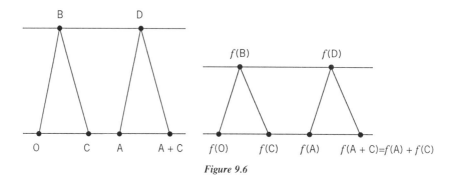

Figure 9.6

$\ell(I_\ell, I_m)\|\ell(A, A')$. It follows from Desargues' Theorem that $\phi_{\ell m} \circ \phi_{k\ell} = \phi_{km}$. The division ring $\mathbb{D}$ is by definition any ring isomorphic to the lines of $\mathcal{L}$.

■ **THEOREM 15.** Suppose that $(\mathcal{P}, \mathcal{L}, \mathcal{E})$ and $(\mathcal{Q}, \mathcal{K}, \mathcal{F})$ are Desarguesian affine spaces, coordinatized by the vector spaces $\mathcal{V}$ and $\mathcal{V}'$ over the division rings $\mathbb{D}$ and $\mathbb{D}'$, and let their lattices of flats be L and L', respectively. If $f : L \to L'$ is a lattice isomorphism, then $f_{|\text{atoms of } L}$ is a semilinear bijection from $\mathcal{V}$ to $\mathcal{V}'$.

Proof: First we show that $\mathbb{D}$ and $\mathbb{D}'$ are isomorphic division rings. Pick O in $\mathcal{P}$, let ℓ be any line containing O, and let I be a second point on ℓ. Then (ℓ, O, I) and $(f(\ell), f(O), f(I))$ are isomorphic division rings. This follows because f, restricted to the atoms of L (the points of $\mathcal{P}$), is a collineation, and as such preserves parallelism. Thus, if A and C are in ℓ, and B $\notin \ell$, $f(A)$ and $f(C)$ are in $f(\ell)$, and $f(B) \notin f(\ell)$. If $\ell(O, B)\|\ell(A, D)$ and $\ell\|\ell(B, D)$, then $\ell(f(O), f(B))\|\ell(f(A), f(D))$ and $f(\ell)\|\ell(f(B), f(D))$. If $\ell(B, C)\|\ell(D, A + C)$, then $\ell(f(B), f(C))\|\ell(f(D), f(A + C))$, as shown in Fig. 9.6. Thus, adding $f(A)$ and $f(C)$ using $f(B)$ and $f(D)$ as the auxiliary points produces $f(A + C)$, so f preserves addition on ℓ. The proof that f preserves multiplication is similar and is left as an exercise.

Since $\mathbb{D} \cong {}_g(\ell, O, I)$ where ℓ is any line of $\mathcal{L}$, and $(f(\ell), f(O), f(I)) \cong {}_h\mathbb{D}'$, then $h \circ f \circ g$ is an isomorphism from $\mathbb{D}$ to $\mathbb{D}'$. This will be denoted σ. If m is any other line through O, $\phi_{f(m)f(\ell)} \circ f \circ \phi_{\ell m} = f$, since f preserves parallelism. Thus, if $a \in \mathbb{D}$, $h \circ \phi_{f(m)f(\ell)} \circ f \circ \phi_{\ell m} \circ g(a) = \sigma(a)$, and the definition of σ does not depend on the choice of ℓ. Furthermore, given $a \in \mathbb{D}$ and any line m containing O, A is the point of m corresponding to the ring element a if and only if $f(A)$ is the point of the line $f(m)$ corresponding to $\sigma(a) \in \mathbb{D}'$.

The restriction S of f to the atoms of L (i.e., to the points in $\mathcal{P}$) preserves vector addition, since f maps parallel lines to parallel lines.

Furthermore $S(a\mathbf{P}) = f(A\mathbf{P})$, where $A = \phi_{\ell m} \circ g(a)$; that is, A is the point corresponding to a in the line $m = \ell(\mathbf{O}, \mathbf{P})$. Thus $f(A\mathbf{P}) = f(A)f(\mathbf{P}) = \sigma(a)f(\mathbf{P})$, since $\sigma(a)f(\mathbf{P})$ is computed on the line $\ell(f(\mathbf{O}), f(\mathbf{P})) = f(\ell(\mathbf{O}, \mathbf{P}))$, so S is a semilinear bijection. Since the affine collineations are transitive on the points of any affine space, $f(\mathbf{O})$ can be taken as the origin of $(\mathcal{D}, \mathcal{K}, \mathcal{F})$ without loss of generality. ∎

Theorem 15 will be used to prove the Fundamental Theorem of Projective Geometry, but first we connect projective collineations with affine ones.

■ **THEOREM 16.** Let $f: L \to L'$ be an isomorphism between the lattices of flats of two projective spaces. If h is a coatom of L, then $f(h)$ is a coatom of L'. Further the restriction of f to $M = \{x \in L : x = 0 \text{ or } x \not\leq h\}$ is an isomorphism from M to $M' = \{y \in L' : y = 0 \text{ or } y \not\leq f(h)\}$.

Proof: By Lemma 1, $f(h)$ is a coatom of L'. Since for all a, b in L, $a \leq b$ if and only if $f(a) \leq f(b)$, it follows that $x \not\leq h$ if and only if $f(x) \not\leq f(h)$, so the restriction of f to M is a bijection from M to M'. Since f and its inverse preserve order in L and L', the same is true for M and M', so $f_{|M}$ is an isomorphism. ∎

The Fundamental Theorem of Projective Geometry, as stated below, essentially says that any vector space over a division ring is completely determined by its lattice of linear subspaces. This is a surprising result in view of the fact that any two groups of prime orders p and q have isomorphic lattices of subgroups, namely the two-element lattice. There are also examples of two groups, one abelian and the other not, that have isomorphic lattices of subgroups. The Fundamental Theorem shows that this cannot happen with vector spaces.

■ **THEOREM 17. The Fundamental Theorem of Projective Geometry.** Let $f: L \to L'$ be an isomorphism between the lattices of subspaces of the vector spaces $\mathcal{V}$ and $\mathcal{V}'$ (of dimension at least 3) over division rings $\mathbb{D}$ and $\mathbb{D}'$. Then f is induced by a semilinear bijection (S, σ). If another semilinear bijection (S_1, σ_1) also induces f, then there is an a in $\mathbb{D}$ such that

$$S_1(\mathbf{x}) = S(a\mathbf{x})$$

and

$$\sigma_1(b) = \sigma(aba^{-1})$$

for all $\mathbf{x}$ in $\mathcal{V}$ and b in $\mathbb{D}$.

Proof: Let h be any coatom of L, and let $M = \{x \in L : x = 0$ or $x \nleq h\}$. Then $f_{|M}$ is an isomorphism from M to $M' = \{y \in L' : y = 0$ or $y \nleq f(h)\}$. But M and M' are lattices of flats of the Desarguesian affine spaces $(\mathscr{P}, \mathscr{L}, \mathscr{E})$ and $(\mathscr{Q}, \mathscr{K}, \mathscr{F})$. Therefore, by Theorem 15, the restriction of f to the atoms of M is a semilinear bijection from the vector space $\mathscr{P}$ over $\mathbb{D}$ to the vector space $\mathscr{Q}$ over $\mathbb{D}'$. Let $\mathscr{V} = \mathscr{P} \times \mathbb{D}$ and $\mathscr{V}' = \mathscr{Q} \times \mathbb{D}'$. Define $S : \mathscr{V} \to \mathscr{V}'$ by $S(\mathbf{x}, a) = (f(\mathbf{x}), \sigma(a))$. Clearly S is a semilinear bijection. If p is an atom of L and $p \nleq h$, then p corresponds to the one-dimensional subspace $\langle (p, 1) \rangle$. $S(\langle (p, 1) \rangle) = \langle (f(p), \sigma(1) \rangle = \langle (f(p), 1) \rangle$ which corresponds to the atom $f(p)$ in L'. If p is an atom of L not under h, then p corresponds to the one-dimensional subspace $\langle (q, 0) \rangle$ where q is a third atom under $p \vee e$ where e is the atom of M that is the zero element of the vector space $\mathscr{P}$. But $f(p) \leq f(q) \vee f(e)$, so $S(\langle (q, 0) \rangle) = \langle (f(q), 0) \rangle$ which corresponds to the atom $f(p)$ in $\mathscr{L}'$.

To complete the proof, suppose that (S, σ) and (S_1, σ_1) both induce f. Then for an arbitrary vector $\mathbf{x} \in \mathscr{V}$, $\langle S(\mathbf{x}) \rangle = \langle S_1(\mathbf{x}) \rangle$, and $S_1(\mathbf{x})$ is a multiple of $S(\mathbf{x})$. Therefore $S_1(\mathbf{x}) = \sigma(a)S(\mathbf{x}) = S(a\mathbf{x})$ for some $a \in \mathbb{D}$. Similarly for $\mathbf{y} \notin \langle \mathbf{x} \rangle$, $S_1(\mathbf{y}) = S(b\mathbf{y})$ and $S_1(\mathbf{x} + \mathbf{y}) = S(c(\mathbf{x} + \mathbf{y}))$. But $S_1(\mathbf{x} + \mathbf{y}) = S_1(\mathbf{x}) + S_1(\mathbf{y}) = S(a\mathbf{x}) + S(b\mathbf{y}) = S(a\mathbf{x} + b\mathbf{y})$; since S is a bijection, $c(\mathbf{x} + \mathbf{y}) = a\mathbf{x} + b\mathbf{y}$. Therefore $a = b = c$, and for any $\mathbf{x}$ in $\mathscr{V}$, $S_1(\mathbf{x}) = S(a\mathbf{x})$ for some fixed $a \in \mathbb{D}$.

For $b \in \mathbb{D}$ and $\mathbf{x}$ in $\mathscr{V}$, $S_1(b\mathbf{x}) = S(ab\mathbf{x}) = \sigma(a)\sigma(b)S(\mathbf{x})$ by the remarks above, but $S_1(b\mathbf{x}) = \sigma_1(b)S_1(\mathbf{x}) = \sigma_1(b)S(a\mathbf{x}) = \sigma_1(b)\sigma(a)S(\mathbf{x})$. Therefore $\sigma(a)\sigma(b) = \sigma_1(b)\sigma(a)$, and $\sigma(aba^{-1}) = \sigma_1(b)$. ∎

For a proof of the Fundamental Theorem that does not use affine geometry, the interested reader can see (Artin 1957, pp. 88–91) and (Jacobson 1974, p. 448).

EXERCISES

1. Let $\{\mathbf{e}_1, \ldots, \mathbf{e}_n\}$ and $\{\mathbf{g}_1, \ldots, \mathbf{g}_n\}$ be bases for the vector spaces $\mathbb{D}^n$ and $\mathbb{D}'^n$, where $\mathbb{D}$ and $\mathbb{D}'$ are division rings, and let σ be an isomorphism from $\mathbb{D}$ to $\mathbb{D}'$. Define $S(a_1\mathbf{e}_1 + \cdots + a_n\mathbf{e}_n) = \sigma(a_1)\mathbf{g}_1 + \cdots + \sigma(a_n)\mathbf{g}_n$. Show that (S, σ) is a semilinear bijection. Show that every semilinear bijection from $\mathbb{D}^n$ to $\mathbb{D}'^n$ can be represented in this way.

2. a. If $(L, \vee, \wedge)$ is the lattice of flats of an affine space, p is an atom of L, and $M = \{x \in L : x \geq p\}$, show that $(M, \vee, \wedge)$ is the lattice of flats of a projective space.

   b. Show that the restriction to M of any lattice isomorphism with domain L is a lattice isomorphism between projective geometries.

3. Prove that the collineations are transitive on the points of any Desarguesian affine space.

9.5 PROJECTIVITIES AND LINEAR TRANSFORMATIONS

By the Fundamental Theorem the automorphisms of any Desarguesian projective space correspond to the semilinear bijections on the corresponding vector space. A natural question is: Which collineations correspond to the *linear* bijections, namely those semilinear bijections (S, σ) where σ is the identity isomorphism? The object of this section is to show that the collineations in question are exactly the projectivities, or finite composites of perspectivities.

Definition. The semilinear bijections (S, σ) and (S_1, σ_1) are said to be *equivalent* if they induce the same collineation between the corresponding projective spaces.

The Fundamental Theorem states that if two semilinear bijections are equivalent, they differ by a scalar. Lemma 4 shows the converse of that statement.

■ **LEMMA 4.** The semilinear bijections (S, σ) and (S_1, σ_1) are equivalent if and only if there is a scalar a such that $S_1(\mathbf{x}) = S(a\mathbf{x})$ and $\sigma_1(b) = \sigma(aba^{-1})$.

Proof: First, it must be shown that S_1 is a semilinear bijection. $S_1(\mathbf{x} + \mathbf{y}) = S(a\mathbf{x} + a\mathbf{y}) = S(a\mathbf{x}) + S(a\mathbf{y}) = S_1(\mathbf{x}) + S_1(\mathbf{y})$, so S_1 preserves vector addition. But

$$\sigma_1(b)S_1(\mathbf{x}) = \sigma(aba^{-1})S(a\mathbf{x}) = \sigma(a)\sigma(b)\sigma(a)^{-1}\sigma(a)S(\mathbf{x})$$

$$= \sigma(a)\sigma(b)S(\mathbf{x}) = S(ab\mathbf{x}) = S_1(b\mathbf{x}),$$

so (S_1, σ_1) is a semilinear bijection.

Since $S_1(\langle \mathbf{x} \rangle) = S(\langle a\mathbf{x} \rangle) = S(\langle \mathbf{x} \rangle)$ for all $\mathbf{x} \in \mathcal{V}$, the effect of S and S_1 is the same on all one-dimensional subspaces of $\mathcal{V}$; thus (S, σ) and (S_1, σ_1) are equivalent.   ■

■ **LEMMA 5.** If an automorphism of a projective space fixes a hyperplane point-wise, it is a perspectivity.

Proof: Suppose that the hyperplane $\mathscr{H}$ is fixed pointwise by the collineation ϕ, and suppose that $\phi(A) \neq A$ (see Fig. 9.7). Let $\angle(A, \phi(A)) \cap \mathscr{H} = X$, and let Y be a point of $\mathscr{H}$ different from X. Let B be any point not in $\mathscr{H}$, not fixed by ϕ, and in the plane α determined by A, X, and Y.

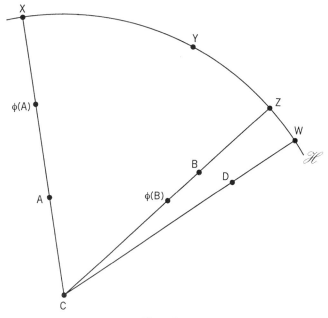

Figure 9.7

Then $\phi(B)$ is also in α, since the lattice isomorphism induced by ϕ maps $\alpha = \{A\} \vee \{X\} \vee \{Y\}$ to $\{\phi(A)\} \vee \{\phi(X)\} \vee \{\phi(Y)\} = \{\phi(A)\} \vee \{X\} \vee \{Y\}$ $= \alpha$. Thus $\ell(A, \phi(A)) \cap \ell(B, \phi(B)) = C$ for some C. Let $\ell(B, \phi(B)) \cap \mathcal{H} = Z$. Then $\phi(C) \in \phi(\ell(A, X)) \cap \phi(\ell(B, Z)) = \ell(A, X) \cap \ell(B, Z) = C$. If $D \neq C$ with $\ell(C, D) \cap \mathcal{H} = W$, then $\phi(D) \in \phi(\ell(C, W)) = \ell(C, W)$. Thus ϕ fixes every line through C and is a perspectivity. ∎

Definition. Given a nonzero element a in a division ring $\mathbb{D}$, the mapping $b \mapsto aba^{-1}$ is called an *inner automorphism* of $\mathbb{D}$.

Thus (S, σ) and (S_1, σ_1) are equivalent semilinear bijections from the vector space $\mathcal{V}$ (over $\mathbb{D}$) to $\mathcal{V}'$ (over $\mathbb{D}'$), if and only if S and S_1 are scalar multiples and $\sigma_1 = \sigma \circ \iota$ where ι is an inner automorphism of $\mathbb{D}$.

■ **LEMMA 6.** Suppose that f is an automorphism of the lattice L of subspaces of a vector space $\mathcal{V}$ (over $\mathbb{D}$) induced by (S, σ). If f fixes all the atoms under a two-dimensional subspace $\mathcal{U}$, then σ is an inner automorphism of $\mathbb{D}$.

Proof: Suppose that $\langle x \rangle$ and $\langle y \rangle$ are one-dimensional subspaces of $\mathcal{U}$ (i.e., atoms under $\mathcal{U}$ in the lattice). Then $\langle S(x) \rangle = \langle x \rangle$ so that $S(x) = ax$ for some $a \in \mathbb{D}$. Similarly $S(y) = by$ and $S(x + y) = c(x + y)$. As in the proof of the last part of the Fundamental Theorem, $a = b = c$, and for every $\mathbf{x} \in \mathcal{U}$, $S(\mathbf{x}) = a\mathbf{x}$. If $a \neq b$ in $\mathbb{D}$, then $S(b\mathbf{x}) = ab\mathbf{x}$ by the remarks above. But $S(b\mathbf{x}) = \sigma(b)S(\mathbf{x}) = \sigma(b)a\mathbf{x}$, so $ab = \sigma(b)a$ and $aba^{-1} = \sigma(b)$. ∎

■ **LEMMA 7.** If σ is an inner automorphism of $\mathbb{D}$, then any semilinear bijection (S, σ) on a vector space $\mathcal{V}$ over $\mathbb{D}$ is equivalent to a linear bijection.

Proof: Let $\sigma(b) = aba^{-1}$, and define (S_1, σ_1) by $S_1(\mathbf{x}) = S(a^{-1}\mathbf{x})$, and $\sigma_1(b) = \sigma(a^{-1}ba)$. By Lemma 4, (S, σ) and (S_1, σ_1) are equivalent semilinear bijections. But $\sigma_1(b) = \sigma(a^{-1}ba) = a(a^{-1}ba)a^{-1} = b$, so σ_1 is the identity and (S_1, σ_1) is a linear bijection. ∎

■ **THEOREM 18.** Every projectivity of any Desarguesian projective space is induced by a linear bijection.

Proof: Suppose that ϕ is a perspectivity on $(\mathscr{P}', \mathscr{L}')$, coordinatized by the vector space $\mathcal{V}$ (of dimension at least 3) over $\mathbb{D}$. By Theorem 12, ϕ fixes a hyperplane pointwise. Thus the lattice automorphism f induced by ϕ fixes a coatom h of the lattice of subspaces of $\mathcal{V}$, and all elements under it. Therefore the semilinear bijection (S, σ) fixes at least a two-dimensional subspace $\mathcal{U}$ of $\mathcal{V}$ and all the one-dimensional subspaces under it. By Lemma 6, σ is an inner automorphism, and by Lemma 7, (S, σ) is equivalent to a linear bijection.

Since the composite of linear bijections is again linear, every projectivity is induced by a linear bijection. ∎

Definition. We will call a linear bijection that fixes every subspace of a hyperplane of a vector space $\mathcal{V}$ an *elementary* linear bijection. Clearly any elementary linear bijection and its inverse fix the same hyperplane; hence the inverse of an elementary linear bijection is elementary as well.

It follows from the proof of Theorem 18 that every perspectivity is induced by an elementary linear bijection. Conversely, every elementary linear bijection on $\mathcal{V}$ fixes a hyperplane and all its subspaces, and therefore by Lemma 5 induces a perspectivity. Thus the elementary linear bijections on $\mathcal{V}$ correspond to the perspectivities of $PG(\mathcal{V})$.

It can be shown that every linear bijection of a finite-dimensional vector space is a finite composite of elementary linear bijections as follows. Any linear bijection S on the vector space $\mathcal{V}$ is completely determined by the bases $\{\mathbf{x}_i\}$ and $\{S(\mathbf{x}_i)\}$. Thus we will use the notation

$$
\begin{pmatrix} \mathbf{x}_1 \\ \mathbf{x}_2 \\ \vdots \\ \mathbf{x}_n \end{pmatrix} \rightarrow \begin{pmatrix} \mathbf{y}_1 \\ \mathbf{y}_2 \\ \vdots \\ \mathbf{y}_n \end{pmatrix} \tag{9.1}
$$

to represent the linear bijection defined by $S(\mathbf{x}_i) = \mathbf{y}_i$ where $\{\mathbf{x}_i\}$ and $\{\mathbf{y}_i\}$ are bases for $\mathscr{V}$. Clearly, if $\{\mathbf{x}_i\}$ and $\{\mathbf{y}_1, \mathbf{x}_2, \ldots, \mathbf{x}_n\}$ are bases for $\mathscr{V}$, then the linear bijection (9.2) below is elementary ($\langle \mathbf{x}_2, \ldots, \mathbf{x}_n \rangle$ being the fixed hyperplane):

$$
\begin{pmatrix} \mathbf{x}_1 \\ \mathbf{x}_2 \\ \vdots \\ \mathbf{x}_n \end{pmatrix} \rightarrow \begin{pmatrix} \mathbf{y}_1 \\ \mathbf{x}_2 \\ \vdots \\ \mathbf{x}_n \end{pmatrix} \tag{9.2}
$$

$$
\begin{pmatrix} \mathbf{x}_1 \\ \mathbf{x}_2 \\ \vdots \\ \mathbf{x}_n \end{pmatrix} \rightarrow \begin{pmatrix} \mathbf{y}_j \\ \mathbf{x}_2 \\ \vdots \\ \mathbf{x}_n \end{pmatrix} \tag{9.3}
$$

Also, given bases $\{\mathbf{x}_i\}$ and $\{\mathbf{y}_i\}$, the linear bijection (9.3) is elementary for some j since $\mathbf{y}_j$ cannot be in $\langle \mathbf{x}_2, \ldots, \mathbf{x}_n \rangle$ for every j. Furthermore the bijection exchanging $\mathbf{x}_1$ and $\mathbf{x}_2$ given by

$$
\begin{pmatrix} \mathbf{x}_1 \\ \mathbf{x}_2 \\ \mathbf{x}_3 \\ \vdots \\ \mathbf{x}_n \end{pmatrix} \rightarrow \begin{pmatrix} \mathbf{x}_1 \\ \mathbf{x}_1 + \mathbf{x}_2 \\ \mathbf{x}_3 \\ \vdots \\ \mathbf{x}_n \end{pmatrix} \rightarrow \begin{pmatrix} \mathbf{x}_2 \\ \mathbf{x}_1 + \mathbf{x}_2 \\ \mathbf{x}_3 \\ \vdots \\ \mathbf{x}_n \end{pmatrix} \rightarrow \begin{pmatrix} \mathbf{x}_2 \\ \mathbf{x}_1 \\ \mathbf{x}_3 \\ \vdots \\ \mathbf{x}_n \end{pmatrix} \tag{9.4}
$$

is a composite of elementary linear bijections. Any linear bijection of the form (9.1) can be achieved by composing linear bijections of the forms (9.2), (9.3), and (9.4). From this it follows:

■ **THEOREM 19.** If $(\mathscr{P}, \mathscr{L})$ is a projective space coordinatized by the finite-dimensional vector space $\mathscr{V}$, then ϕ is a projectivity on $(\mathscr{P}, \mathscr{L})$ if and only if it is induced by a linear bijection on $\mathscr{V}$.

EXERCISES

1. a. Show that $\{(3, 0, 1), (1, 2, 1), (0, 0, 1)\}$ is a basis for $\mathbb{R}^3$.
   b. Express the linear bijection $T : \mathbb{R}^3 \rightarrow \mathbb{R}^3$ given by $T(x, y, z) = (3x + y, 2y, x + y + z)$ as a composite of elementary linear bijections.

2. Define $T : (\mathbb{Z}_5)^4 \rightarrow (\mathbb{Z}_5)^4$ by $T(w, x, y, z) = (w + x, x + y + z, z + y, y)$. Express T as a composite of elementary linear bijections.

3. Suppose that T is a linear bijection of a vector space $\mathcal{V}$ over a field F. Recall that $\mathbf{x} \in \mathcal{V}$ is an *eigenvector* of T, with *eigenvalue* $a \in F$ if $T(\mathbf{x}) = a\mathbf{x}$. Interpret the concepts of eigenvector and eigenvalue geometrically (i.e., in $AG(\mathcal{V})$ and $PG(\mathcal{V})$). How does the situation differ if F is a noncommutative division ring?

4. Suppose that T is a linear bijection of a vector space $\mathcal{V}$ over a field F. Suppose that $\mathbf{x}$, $\mathbf{y}$, and $\mathbf{z}$ are eigenvectors for T with distinct eigenvalues a, b, and c, respectively. Show that $\mathbf{x}$, $\mathbf{y}$, and $\mathbf{z}$ are linearly independent.

*5. Suppose that $f : M \to M$ is an automorphism on the lattice of affine flats of a vector space $\mathcal{V}$ over a field F induced by a linear transformation T. Show that T has an eigenvector with a nonzero eigenvalue if and only if there is an atom p and a line ℓ both fixed by f. [*Hint*: Coordinatize by taking p to be the zero vector, and consider the cases where $p \leq \ell$ and $p \not\leq \ell$.]

6. Use the fact that the only isomorphism of the real numbers is the identity to show that the only collineations of $PG(\mathbb{R}^n)$ are the projectivities.

7. Define an *affine homomorphism* to be a mapping (not necessarily a bijection) $\psi : (\mathcal{P}, \mathcal{L}, \mathcal{E}) \to (\mathcal{Q}, \mathcal{K}, \mathcal{F})$ that satisfies the condition $\psi(\ell(A, B)) = \ell(\psi(A), \psi(B))$ if $\psi(A) \neq \psi(B)$, and $\psi(\ell(A, B)) = \psi(A)$ if $\psi(A) = \psi(B)$.

   a. If $\alpha = \overline{\{A, B, C\}}$, show that $\psi(\alpha) \subseteq \overline{\{\psi(A), \psi(B), \psi(C)\}}$.

   b. If $\ell \| m$ in $\mathcal{L}$, show that $\psi(\ell)$ is parallel or equal to $\psi(m)$ in $\mathcal{K}$.

   c. If the image of ψ is not contained in a single line, let $\psi(\ell) = \ell' \in \mathcal{K}$ for some $\ell \in \mathcal{L}$. Show that for any choice of O and I in ℓ, the division rings $(\ell, \text{O}, \text{I})$ and $(\ell', \psi(\text{O}), \psi(\text{I}))$ are isomorphic.

   *d. Show that ψ is induced by a semilinear transformation (S, σ), namely a mapping, not necessarily a bijection, $S : \mathcal{P} \to \mathcal{Q}$, that preserves vector addition and such that $S(a\text{P}) = \sigma(a)S(\text{P})$, where σ is an isomorphism between the coordinatizing division rings of $(\mathcal{P}, \mathcal{L}, \mathcal{E})$ and $(\mathcal{Q}, \mathcal{K}, \mathcal{F})$.

   e. Show that ψ induces a lattice homomorphism f (a join and meet preserving function) from the lattice of flats of $(\mathcal{P}, \mathcal{L}, \mathcal{E})$ to the lattice of flats of $(\mathcal{Q}, \mathcal{K}, \mathcal{F})$.

9.6 COLLINEATIONS AND COMMUTATIVITY

It is well known that any linear bijection is completely determined by its action on a basis of a vector space. However, the same is not true for projective collineations, as the following example shows:

Suppose that $T : \mathbb{R}^3 \to \mathbb{R}^3$ is given by

$$T(1, 0, 0) = (2, 0, 0),$$

$$T(0, 1, 0) = (0, 1, 0),$$

and

$$T(0, 0, 1) = (0, 0, 1).$$

The projectivity induced by T is not the identity, since it maps $\langle 1, 1, 0 \rangle$ to $\langle 2, 1, 0 \rangle$. However, it agrees with the identity on the projective points $\langle 1, 0, 0 \rangle \ (= \langle 2, 0, 0 \rangle)$, $\langle 0, 1, 0 \rangle$, and $\langle 0, 0, 1 \rangle$. Thus more than three noncollinear projective points are needed to determine a projectivity of $PG(\mathbb{R}^3)$.

It will be shown in this section that $n + 1$ points and their images determine a unique projectivity on $PG(\mathbb{D}^n)$ if and only if $\mathbb{D}$ is commutative.

The following result is called the *Second Fundamental Theorem of Projective Geometry* in (Baer 1952, p. 66) and is stated as a corollary to the Fundamental Theorem of Projective Geometry in (Artin 1957, p. 98).

■ **THEOREM 20.** Let $\mathscr{V}$ be a vector space with bases $\{\mathbf{x}_1, \mathbf{x}_2, \ldots, \mathbf{x}_n\}$ and $\{\mathbf{y}_1, \mathbf{y}_2, \ldots, \mathbf{y}_n\}$ $(n \geq 3)$ over a division ring $\mathbb{D}$. The following are equivalent:

   i. $\mathbb{D}$ is commutative.
   ii. If a projectivity ϕ fixes $\langle \mathbf{x}_i \rangle$ $(i = 1, \ldots, n)$ and one additional subspace $\langle \mathbf{x}_{i+1} \rangle$, ϕ is the identity.
   iii. Given the additional one-dimensional subspaces $\langle \mathbf{x}_{n+1} \rangle$ and $\langle \mathbf{y}_{n+1} \rangle$, such that no proper subset of $\{\mathbf{x}_1, \ldots, \mathbf{x}_{n+1}\}$, is linearly dependent and no proper subset of $\{\mathbf{y}_1, \ldots, \mathbf{y}_{n+1}\}$ is linearly dependent, there is a unique projectivity ϕ such that $\phi(\langle \mathbf{x}_i \rangle) = \langle \mathbf{y}_i \rangle$ for $i = 1, \ldots, n + 1$.

Proof: (This proof is adapted from the proof in Artin 1957, p. 98.)

i → ii: If $\{\mathbf{x}_i\}_{i=1,\ldots,n}$ is a basis for $\mathscr{V}$, so is $\{a_i \mathbf{x}_i\}_{i=1,\ldots,n}$ for nonzero scalars a_i. Since $\mathbf{x}_{n+1} = a_1 \mathbf{x}_1 + \cdots + a_n \mathbf{x}_n$ (a_i nonzero) and since $\langle \mathbf{x}_i \rangle = \langle a_i \mathbf{x}_i \rangle$, one can replace $\mathbf{x}_i$ by $a_i \mathbf{x}_i$ and assume that $\mathbf{x}_{n+1} = \mathbf{x}_1 + \cdots + \mathbf{x}_n$. If T is the linear transformation induced by ϕ, then $T(\mathbf{x}_i) = b_i \mathbf{x}_i$ for some b_i $(i = 1, \ldots, n)$. Since T is linear, $T(\mathbf{x}_{n+1}) = T(\mathbf{x}_1 + \cdots + \mathbf{x}_n) = b_1 \mathbf{x}_1 + \cdots + b_n \mathbf{x}_n$. But T fixes $\langle \mathbf{x}_{n+1} \rangle$, so $T(\mathbf{x}_{n+1}) = c \mathbf{x}_{n=1}$ for some scalar c. Thus $b_1 \mathbf{x}_1 + \cdots + b_n \mathbf{x}_n = c(\mathbf{x}_1 + \cdots + \mathbf{x}_n)$, the b_i are all equal to c, and the transformation T is simply scalar multiplication by c, since $T(a_1 \mathbf{x}_1 + \cdots + a_n \mathbf{x}_n) = a_1 T(\mathbf{x}_1) + \cdots + a_n T(\mathbf{x}_n) = a_1 c \mathbf{x}_1 + \cdots + a_n c \mathbf{x}_n$, which equals $c(a_1 \mathbf{x}_1 + \cdots + a_n \mathbf{x}_n)$ since $\mathbb{D}$ is commutative. Thus $T(\langle \mathbf{x} \rangle) = \langle c \mathbf{x} \rangle = \langle \mathbf{x} \rangle$ for every $\mathbf{x} \in \mathscr{V}$. Therefore T maps every subspace of $\mathscr{V}$ to itself, and the projectivity ϕ is the identity collineation.

ii → i: Suppose that $ab \neq ba$ in $\mathbb{D}$. Define $T(\mathbf{x}_i) = a \mathbf{x}_i$. Thus $T(\mathbf{x}_1 + \cdots + \mathbf{x}_n) = a(\mathbf{x}_1 + \cdots + \mathbf{x}_n)$, and the projectivity ϕ that induces T satisfies the hypotheses of ii. But $T(\mathbf{x}_1 + b\mathbf{x}_2) = T(\mathbf{x}_1) + bT(\mathbf{x}_2) = a\mathbf{x}_1 + ba\mathbf{x}_2 \neq a\mathbf{x}_1 + ab\mathbf{x}_2 = a(\mathbf{x}_1 + b\mathbf{x}_2)$. Thus T does not fix the subspace $\langle \mathbf{x}_1 + b\mathbf{x}_2 \rangle$, so ϕ is not the identity collineation.

ii → iii: Assuming as above that $\mathbf{x}_{n+1} = \mathbf{x}_1 + \cdots + \mathbf{x}_n$, and similarly for the $\mathbf{y}_i$, define $T(\mathbf{x}_i) = \mathbf{y}_i$, $i = 1, \ldots, n$. Then $T(\mathbf{x}_{n+1}) = \mathbf{y}_{n+1}$, and the projectivity corresponding to T satisfies the conditions for iii. Suppose that ϕ and ψ map $\langle \mathbf{x}_i \rangle$ to $\langle \mathbf{y}_i \rangle$, $i = 1, \ldots, n + 1$. Then $\phi^{-1} \circ \psi$ is a projectivity on $\mathcal{V}$ that fixes $\langle \mathbf{x}_i \rangle$ for $i = 1, \ldots, n + 1$ and is the identity. Thus $\phi = \psi$.

iii → ii: The proof follows trivially by setting $\mathbf{y}_i = \mathbf{x}_i$ for $i = 1, \ldots, n + 1$. ■

Since Pappus' Theorem holds in a projective space if and only if the coordinatizing division ring is commutative, the following corollary is immediate:

Corollary. Pappus' Theorem holds in $PG(\mathcal{V})$ if and only if any (and therefore all) of the conditions in Theorem 20 hold.

EXERCISES

1. Describe two collineations which fix $\langle 1, 1, 1 \rangle$, $\langle 3, 1, 2 \rangle$, and $\langle 4, 2, 1 \rangle$ in $P_2(\mathbb{Z}_5)$. Describe the only collineation that fixes $\langle 3, 4, 4 \rangle$ in addition to the other three projective points.
2. Describe two collineations of the real projective plane that map $\langle 3, 0, 1 \rangle$ to $\langle 0, 0, 2 \rangle$, $\langle 1, 2, 1 \rangle$ to $\langle 4, 0, 1 \rangle$, and $\langle 0, 0, 1 \rangle$ to $\langle 1, 3, 1 \rangle$. Describe the unique collineation that also maps $\langle 4, 2, 3 \rangle$ to $\langle 5, 3, 4 \rangle$.

SUGGESTED READING

Emil Artin, *Geometric Algebra*, New York: Wiley Interscience, 1957.

Reinhold Baer, *Linear Algebra and Projective Geometry*, San Diego: Academic Press, 1952.

Marshall Hall, Projective planes, *Trans. AMS* **54** (1943), 229–277.

Daniel R. Hughes and Fred C. Piper, *Projective Planes*, New York: Springer-Verlag, 1972.

Nathan Jacobson, *Basic Algebra I*, San Francisco: Freeman, 1974.

Frederick Stevenson, *Projective Planes*, San Francisco: Freeman, 1972. Reprint, Washington, NJ: Polygonal Press, 1992.

Algebraic Background

This appendix gives an introduction to the theory of fields, particularly finite fields, providing enough information for the reader to construct some small finite fields, which can be used to give examples of finite affine and projective spaces.

A.1 FIELDS

In this section we repeat the definition of a field given in Chapter 1 and discuss the basic properties of fields. A more detailed treatment can be found in any abstract algebra text, including those listed in the references at the end of this appendix.

Definition. A *field* is a set F with two operations, addition and multiplication, satisfying the following properties for all a, b, and c in F:

$F1$. $a + b$ is in F whenever a and b are in F (closure).

$F2$. $a + b = b + a$ (commutativity).

$F3$. $a + (b + c) = (a + b) + c$ (associativity).

$F4$. There is an element 0 (called the *additive identity*) in F with $0 + a = a$ (existence of additive identity).

$F5$. For each a in F there is an element $-a$ in F with $a + (-a) = 0$ (existence of additive inverses).

$F6$. ab is in F whenever a and b are in F (closure).

$F7$. $ab = ba$ (commutativity).

$F8$. $a(bc) = (ab)c$ (associativity).

$F9$. There is an element 1 ($\neq 0$) in F (called its *multiplicative identity*) with $1a = a$ (existence of multiplicative identity).

F10. If $a \neq 0$, there is an element a^{-1} in F with $aa^{-1} = 1$ (existence of multiplicative inverses)

F11. $a(b + c) = ab + ac$; $(b + c)a = ba + ca$ (distributivity).

Properties F1–F5 define F as an *abelian group* under addition. The rational numbers and the real numbers are fields. Less obviously the complex numbers also form a field as seen in the following example.

□**EXAMPLE 1:** The complex numbers of the form $a + bi$, a and b real, where $(a + bi) + (c + di) = (a + c) + (b + d)i$ and $(a + bi) \times (c + di) = (ac - bd) + (ad + bc)i$ form a field $(\mathbb{C}, +, \times)$ in which $i^2 = -1$.   □

■ **THEOREM 1.** (*The cancellation law for addition*). In any field F, if $a + b = a + c$, then $b = c$. If $b + a = c + a$, then $b = c$.

Proof: If $a + b = a + c$, then $(-a) + (a + b) = (-a) + (a + c)$, so $[(-a) + a] + b = [(-a) + a] + c$; thus $0 + b = 0 + c$ and $b = c$. The second statement follows from the commutativity of addition.   ■

Corollary. For any element a contained in a field F, $a = -(-a)$.

Proof: By definition of $-(-a)$, $-a + -(-a) = 0$. By the cancellation law, since $-a + a = 0$, $a = -(-a)$.   ■

■ **THEOREM 2.** The following statements are true for every a and b in any field F:

   i. $a0 = 0$.
   ii. $0a = 0$.
   iii. $a(-b) = -(ab)$.
   iv. $(-a)b = -(ab)$.
   v. $(-a)(-b) = ab$.

Proof: (i) $a0 = a(0 + 0) = a0 + a0$. But $a0 + 0 = a0$. Thus by the cancellation law for addition, $a0 = 0$. Part ii follows from commutativity of multiplication. (iii) $a(-b) + ab = a(-b + b) = a0 = 0$. Thus $a(-b) = -(ab)$. Part iv follows from commutativity of multiplication. (v) $(-a)(-b) = -(a(-b))$ by part iv, and $-(a(-b)) = -(-(ab))$ by part iii. By the corollary to the cancellation law, $-(-(ab)) = ab$.   ■

The theorem above gives the reason why only the nonzero field elements have multiplicative inverses. If there were an element 0^{-1}, then by F10, $00^{-1} = 1$, while by part ii of Theorem 2, $00^{-1} = 0$. Since $1 \neq 0$ by F9, 0 cannot have a multiplicative inverse.

EXERCISES

1. Show that the complex number $1 + 0i$ is the *multiplicative identity* for $\mathbb{C}$, and that for a and b not both 0, $a + bi$ has a unique *multiplicative inverse*.

2. Prove the multiplicative cancellation laws for fields. That is, if $a \times b = a \times c$ (or $b \times a = c \times a$) *and a is different from* 0, then $b = c$.

3. Prove that the additive identity in any field is unique and that the multiplicative identity in any field is unique.

4. Show that the multiplicative inverse of any nonzero field element is unique. [*Hint*: Use the multiplicative cancellation law.]

5. Prove that, in any field F, if $a \times b = 0$, then $a = 0$ or $b = 0$.

6. Prove that $\mathbb{Q}(\sqrt{2}) = \{a + b\sqrt{2} : a, b \in \mathbb{Q}\}$ is a field under real number addition and multiplication.

7. Construct addition and multiplication tables for a field containing four elements, $0, 1, a, b$. [*Hint*: $x + x = 0$ for all x in this field.]

8. Solve the equation $ay + b = 1$ for y in the four element field found in Exercise 7.

A.2 THE INTEGERS MOD n

A set with binary operations addition and multiplication that satisfies all of the field properties except $F10$ is called a *commutative ring*. The most familiar example is the ring of integers $\mathbb{Z}$ with the usual addition and multiplication. Both these operations are associative and commutative, 0 and 1 are the identities, $-a$ is the additive inverse for any integer a, and the distributive laws hold. The only integers that have multiplicative inverses are 1 and -1; hence the integers do not form a field. The systems $(\mathbb{Z}_n, +_n, \times_n)$ introduced informally in Section 2.3 are always commutative rings, and are fields if and only if n is a prime number.

Recall that given a positive integer n, the integers x and y are said to be *congruent mod n* (written $x \equiv_n y$) when $(x - y)$ is divisible by n. Furthermore, given any positive integers x and n, there is a unique k such that

$$x \equiv_n k, \qquad 0 \le k \le n - 1.$$

The number k is the remainder obtained upon dividing x by n. Thus the equivalence classes relative to $\equiv_n$ can be identified with the set of residues $\mathbb{Z}_n = \{0, 1, \dots, n - 1\}$.

Addition and multiplication can be defined in $\mathbb{Z}_n$ as follows:

$$a +_n b = k, \qquad k \in \mathbb{Z}_n, \qquad a + b \equiv k \pmod{n},$$

and

$$a \times_n b = m, \qquad m \in \mathbb{Z}_n, \qquad a \times b \equiv m \pmod{n}.$$

Alternatively, $a +_n b$ is the remainder obtained when the usual sum $a + b$ is divided by n, and similarly for multiplication.

□**EXAMPLE 2:** For any integer $n \geq 2$, $(\mathbb{Z}_n, +_n, \times_n)$ is a commutative ring. The closure properties follow from the discussion above, while commutativity, associativity and distributivity hold because these are properties of the integers. The elements 0 and 1 are the additive and multiplicative identities, respectively, and the additive inverse of any nonzero element k in $\mathbb{Z}_n$ is $n - k$, with 0 being its own additive inverse.   □

□**EXAMPLE 3:** From the addition and multiplication tables for $\mathbb{Z}_3$, it is apparent that 1 and 2 are their own multiplicative inverses. Therefore $\mathbb{Z}_3$ is a field, whose tables are

+	0	1	2		×	0	1	2
0	0	1	2		0	0	0	0
1	1	2	0		1	0	1	2
2	2	0	1		2	0	2	1

Tables for $\mathbb{Z}_3$ □

As a specific example of a $\mathbb{Z}_n$ that is not a field, consider the multiplication table for $\mathbb{Z}_6$, and observe that only 1 and 5 have multiplicative inverses:

$\times_6$	0	1	2	3	4	5
0	0	0	0	0	0	0
1	0	1	2	3	4	5
2	0	2	4	0	2	4
3	0	3	0	3	0	3
4	0	4	2	0	4	2
5	0	5	4	3	2	1

Multiplication Table for $\mathbb{Z}_6$ □

The rings $\mathbb{Z}_n$ are commutative rings; they are fields if and only if each nonzero element has a multiplicative inverse. This depends on n, as the following theorem shows:

■ **THEOREM 3.** $\mathbb{Z}_n$ is a field if and only if n is a prime number.

Proof: Suppose that $\mathbb{Z}_n$ is a field, and that $n = pq$, where p and q are between 2 and $n - 1$ (inclusive). Let p^{-1} be the inverse of p in $\mathbb{Z}_n$. Then $(p^{-1} \times_n p) \times_n q = q$, but the associative law implies that $p^{-1} \times_n (p \times_n q) = p^{-1} \times_n 0 = 0$, so $q = 0$, a contradiction.

Conversely, if n is prime, and p is a nonzero member of $\mathbb{Z}_n$, consider the products $0 \times_n p, 1 \times_n p, \ldots, (n-1) \times_n p$. If none of these is equal to 1, then two of them must be equal to each other. Suppose that $i \times_n p = (i+j) \times_n p$ where j is different from 0. Then by distributivity $j \times_n p = 0$. But this means that in ordinary multiplication, jp is either 0 or n. In the first instance j or p is zero; in the second, n is not prime. In either event there is a contradiction; hence $i \times_n p = 1$ for some i. ∎

Definition. The fields F and F' are *isomorphic* (denoted $F \cong F'$) if there is a bijection $\phi : F \to F'$ satisfying the following conditions for all $a, b \in F$:

1. $\phi(a+b) = \phi(a) + \phi(b)$.
2. $\phi(ab) = \phi(a)\phi(b)$.

EXERCISES

1. Make addition and multiplication tables for $\mathbb{Z}_4$ and $\mathbb{Z}_5$, noting that $\mathbb{Z}_5$ is a field while $\mathbb{Z}_4$ is not.

2. Prove that an element $k \in \mathbb{Z}_n$ has a multiplicative inverse in $\mathbb{Z}_n$ if and only if k and n are relatively prime.

3. When possible, solve the following equations:
   a. $2x + 4 = 1$ in $\mathbb{Z}_5$.
   b. $3x + 5 = 2$ in $\mathbb{Z}_7$.
   c. $2x = 1$ in $\mathbb{Z}_4$.
   d. $x^2 = 2$ in $\mathbb{Z}_7$.
   e. $x^2 = 3$ in $\mathbb{Z}_7$.
   f. $x^2 + x + 1 = 0$ in $\mathbb{Z}_2$.

4. Solve the following simultaneous equations in $\mathbb{Z}_7$:

$$\begin{cases} 2x + 5y = 1, \\ 3x + 1y = 6. \end{cases}$$

5. a. Prove that the composite of field isomorphisms is again a field isomorphism.
   b. Prove that any field isomorphism maps 0 to 0 and 1 to 1.
   c. Prove that the four-element field constructed in Exercise 7 of Section A.1 is not isomorphic to $\mathbb{Z}_4$.

6. Prove that the field of real numbers is not isomorphic to the field of complex numbers.

*7. A commutative ring is called an *integral domain* if it satisfies the following cancellation law for multiplication: if $a \times b = a \times c$ and $a \neq 0$, then $b = c$. Prove that any finite integral domain is a field.

*8. Let $\mathscr{X}$ by any nonempty set, and let $2^{\mathscr{X}}$ be the collection of all its subsets. Prove that $2^{\mathscr{X}}$ is a commutative ring where $A + B = (A \cap B') \cup (A' \cap B)$ and $A \times B = A \cap B$ (A and B subsets of $\mathscr{X}$). Is $2^{\mathscr{X}}$ a field? Is it an integral domain? (By A' is meant the *set complement* of A, namely $\mathscr{X} \setminus A$).

A.3 FINITE FIELDS

Recall that the complex numbers are of the form $a + bi$, where a and b are real numbers and $i^2 = -1$, or equivalently that i satisfies the equation $x^2 + 1 = 0$, an equation with no real solutions. These binomials are added and multiplied as polynomials in i subject to the rule that $i^2 = -1$.

Analogously, the fields $\mathbb{Z}_p$ can be used to define larger finite fields in a similar way. Since the variables x and y are used for linear equations, we will use u as the variable in constructing finite fields. The smallest such example follows.

☐**EXAMPLE 4:** A four-element field can be formed from $\mathbb{Z}_2$ as the set of elements $a + bu$, where a and b are in $\mathbb{Z}_2$ and u satisfies the equation $u^2 = u + 1$, the calculation being done for polynomials with coefficients in $\mathbb{Z}_2$, subject to the rule that $u^2 = u + 1$. This field is called $GF(4)$ [the initials standing for "Galois Field"], and its elements are $0, 1, u$, and $u + 1$. For example, adding the elements u and $u + 1$ gives $u + u + 1 = (1 + 1)u + 1 = 0u + 1 = 1$. Multiplying u by $u + 1$ gives $u^2 + u = (u + 1) + u = 1$. Note that this field is different from $\mathbb{Z}_4$, since for every element $y \in GF(4)$, $y + y = 0$, while in $\mathbb{Z}_4$, $3 + 3 = 2 \neq 0$.   ☐

The field above has $4 = 2^2$ elements, and it was based on $\mathbb{Z}_2$. In fact for every prime number p and positive integer k there is one and only one field (up to isomorphism) with p^k elements, denoted $GF(p^k)$. Furthermore every finite field has p^k elements for some prime p and positive integer k. Here we will show how to construct some fields of order p^2 and p^3.

Definition. If F is a field, we denote by $F[u]$ the set of polynomials in u with coefficients in F, namely $q(u) \in F[u]$ if

$$q(u) = a_0 + a_1 u + a_2 u^2 + \cdots + a_n u^n,$$

where the a_i are elements of F, and u is an indeterminate. The highest exponent (here n) is called the *degree* of the polynomial. The polynomials in $F[u]$ can be added and multiplied by the same rules as for real polynomials; however, the computations are performed in F.

For example, the sum of the polynomials $4u^2 + 2u + 1$ and $3u^2 + u + 4$ in $\mathbb{Z}_5[u]$ is $2u^2 + 3u$, while their product is $2u^4 + u^2 + 4u + 4$.

If $q(u)$ cannot be factored into polynomials of strictly smaller degree in $F[u]$, then $q(u)$ is said to be *irreducible over F*, and $q(u)$ can be used to construct a field from $F[u]$. We will do this only for polynomials of degree 2 and 3 since, for a general higher-degree polynomial, there is usually no easy way to tell when it can be factored. (See Lidl and Pilz 1984, pp. 163–166, for tables of irreducible polynomials of higher degrees over various $\mathbb{Z}_p$.) However, if p has degree 2 or 3, and can be factored into polynomials of strictly smaller degree, then one of the factors has degree 1. That factor is of the form $au - b$, so $u = a^{-1}b$ is a *root* of $q(u)$, and a solution of $q(u) = 0$.

Definition. For $q(u) = u^2 - ru - s$ an irreducible polynomial in $F[u]$, the field $F[u]/(q(u)) = \{a + bu : a, b \in F\}$ where addition is the addition of $F[u]$ and multiplication is the same as in $F[u]$, except that u^2 is equal to $ru + s$.

It follows that the complex field $\mathbb{C}$ is isomorphic to $\mathbb{R}[u]/(u^2 + 1)$, and the four-element field $GF(4) = \mathbb{Z}_2[u]/(u^2 + u + 1)$. Since $\sqrt[2]{2}$ is irrational, the polynomial $u^2 - 2$ has no roots in $\mathbb{Q}$. Consequently the field $\mathbb{Q}[u]/(u^2 - 2)$ has as its elements $\{a + bu : a, b \in \mathbb{Q}\}$ where $u^2 = 2$.

□**EXAMPLE 5: The field $GF(9)$.** To construct a nine-element field $FG(9)$, we must find a polynomial of degree 2 with coefficients in $\mathbb{Z}_3$ that has no roots in $\mathbb{Z}_3$. One example is $q(u) = u^2 - u - 1$. Note that $q(0) = -1$, $q(1) = -1 (= 2)$, and $q(2) = 1$. Thus $GF(9) = \{au + b : a, b \in \mathbb{Z}_3\}$, where $u^2 = u + 1$. The sum of $2u + 1$ and $2u + 2$ is u, while their product is $(2u + 1)(2u + 2) = u^2 + 2u + u + 2 = u + 1 + 2 = u$. □

Definition. For $q(u) = u^3 - ru^2 - su - t$ an irreducible polynomial in $F[u]$, $F[u]/(q(u)) = \{a + bu + cu^2 : a, b, c \in F\}$ is a field under polynomial addition and multiplication in $F[u]$, with u^3 equal to $ru^2 + su + t$.

Thus $u^4 = u^3 u = (ru^2 + su + t)u = r(ru^2 + su + t) + su^2 + tu = (r^2 + s)u^2 + (rs + t)u + rt$.

□**EXAMPLE 6: The field $GF(27)$.** The field with 27 elements can be constructed from $\mathbb{Z}_3$ by using the polynomial $u^3 + 2u^2 + 1 = u^3 - u^2 - 2$, which has neither 0, 1, nor 2 as a root and where u^3 is set equal to $u^2 + 2$. □

The Characteristic of a Field

In $\mathbb{Z}_2$, $1 + 1 = 0$, while in $\mathbb{Z}_3$, $1 + 1 + 1 = 2 + 2 + 2 = 0$. In a general field the smallest positive integer n such that na, which is defined to be $a + a + \cdots + a$ (a written n times), is zero for every a in the field is called the *characteristic* of the field. If no such n exists, the field is said to have characteristic 0 (some books say characteristic ∞). Since $a + a + \cdots + a = a(1 + 1 + \cdots + 1)$, then the characteristic of a field F is n exactly when n is the smallest positive integer such that $n1 = 0$. From this it follows that the characteristic of $\mathbb{Z}_p$ is p. Furthermore, since $GF(4)$ consists of elements of the form $b = a_1 + a_2 u(a_i = 0$ or 1) which are added as in $\mathbb{Z}_p$, then $1 + 1 = 0$, and the characteristic of $GF(4)$ is 2. Similarly it follows that the characteristic of any $\mathbb{Z}_p[u]/(q(u))$ is p.

EXERCISES

1. Compute the $f(u) + g(u)$ and $f(u)g(u)$ for the following pairs of polynomials:
   a. $u^2 + 2u + 1$ and $u^3 + u + 2$ in $\mathbb{Z}_3[u]$.
   b. $u^2 + 4u + 2$ and $u^4 + u$ in $\mathbb{Z}_5[u]$.
   c. $u^3 + u + 1$ and $u^2 + 4u + 2$ in $\mathbb{Z}_7[u]$.

2. a. Construct addition and multiplication tables for the nine-element field $\mathbb{Z}_3[u]/(u^2 + 1)$.
   *b. SHow that the field in (a) is isomorphic to $\mathbb{Z}_3[u]/(u^2 - u - 1)$.
   *c. Given that $u^4 + u + 1$ is irreducible over $\mathbb{Z}_2$, describe the field $GF(16)$.
   *d. Given that $u^4 - u^2 - 1$ is irreducible over $\mathbb{Z}_3$, describe the field $GF(81)$.

3. In $\mathbb{Z}_3[u]/(u^2 - u - 1)$, show that the elements $u, u^2, u^3, \ldots, u^7, u^8$ take on all nonzero values in the field, with $u^8 = 1$.

4. a. Find an irreducible polynomial of degree 3 over $\mathbb{Z}_5$, and use it to describe a field with 125 elements.
   b. Describe the fields $GF(49)$ and $GF(343)$.

5. Prove that the characteristic of any field is either 0 or a prime number. [*Hint:* See Exercise 5 of Section A.1.]

6. In this exercise, refer to the addition and multiplication tables given for $GF(8)$. Use the tables to solve these equations for x and y in $GF(8)$:
   a. $px + t = v$.
   b. $\left.\begin{matrix} qx + ty = r \\ x + py = s \end{matrix}\right\}$ [*Hint:* Solve these as simultaneous equations.]

+	0	1	p	q	r	s	t	v
0	0	1	p	q	r	s	t	v
1	1	0	s	t	v	p	q	r
p	p	s	0	r	q	1	v	t
q	q	t	r	0	p	v	1	s
r	r	v	q	p	0	t	s	1
s	s	p	1	v	t	0	r	q
t	t	q	v	1	s	r	0	p
v	v	r	t	s	1	q	p	0

×	0	1	p	q	r	s	t	v
0	0	0	0	0	0	0	0	0
1	0	1	p	q	r	s	t	v
p	0	p	q	s	v	r	1	t
q	0	q	s	r	t	v	p	1
r	0	r	v	t	p	1	s	q
s	0	s	r	v	1	t	q	p
t	0	t	1	p	s	q	v	r
v	0	v	t	1	q	p	r	s

Tables for $GF(8)$ as an abstract field

7. a. Show that $GF(8) \cong \mathbb{Z}_2[u]/(u^3 + u + 1)$ as follows: Check that the element p satisfies the equation $u^3 = u + 1$.

  b. Write the elements q, r, s, t, and v in the form $a_0 + a_1 p + a_2 p^2$ where a_0, a_1, and a_2 are in $\mathbb{Z}_2$ (i.e., they are either 0 or 1).

  c. Show that $p, p^2, \ldots, p^7$ are all distinct with $p^7 = 1$.

SUGGESTED READING

G. Birkhoff and S. Mac Lane, *A Survey of Modern Algebra*, 4th ed., New York: Macmillan, 1977.

R. Lidl and G. Pilz, *Applied Abstract Algebra*, New York: Springer-Verlag, 1984.

Hilbert's Example of a Noncommutative Division Ring

In his *Grundlagen der Geometrie* (translated as *Foundations of Geometry*), David Hilbert gave a complete set of axioms for Euclidean geometry, the first three groups of which were the incidence, betweenness (order), and parallel axioms. The incidence and parallel axioms give an affine geometry over a division ring $\mathbb{D}$, and the order axioms imply that the ring is *ordered* in that there is a reflexive antisymmetric and transitive relation $\leq$ on $\mathbb{D}$ such that $a \leq b$ or $b \leq a$ whenever a and b are in $\mathbb{D}$. Furthermore, if $a \leq b$, then $a + c \leq b + c$ for any c, and $ac \leq bc$ if $0 \leq c$.

To present an example of an affine plane $\mathbb{D}^2$ in which the incidence, order, and parallel axioms hold but Pappus' Theorem does not, Hilbert gave the following example of a noncommutative division ring that can be ordered (see Hilbert 1902, sec. 33).

First consider the *formal Laurent series*, namely the (possibly infinite) polynomial expressions of the form

$$T = r_0 x^n + r_1 x^{n+1} + r_2 x^{n+2} + r_3 x^{n+3} + \cdots, \qquad (B.1)$$

where the coefficients r_i are rational numbers and the exponent n is a (possibly negative) integer. Now look at expressions of the form

$$S = y^m T_0 + y^{m+1} T_1 + y^{m+2} T_2 + \cdots, \qquad (B.2)$$

where the T_i are expressions of the form given in (B.1), and m is a (possibly negative) integer. $\mathbb{D}$ consists of all power series of the form (B.2).

Addition in $\mathbb{D}$ is performed as for polynomials over $\mathbb{Q}$. For example, if

$$S_1 = y^{-2}(x^{-3} + 2x^2) + y^2(x^{-1} + 3x^2 + 4x^3)$$

and

$$S_2 = y^{-2}(x^{-4} + 2x^{-3} + x^2) + y^{-1}x + y^2(x^2 + x^3),$$

then

$$S_1 + S_2 = y^{-2}(x^{-4} + 3x^{-3} + 3x^2) + y^{-1}x + y^2(x^{-1} + 4x^2 + 5x^3).$$

Multiplication is done as for polynomials, except that

$$xy = 2yx. \tag{B.3}$$

The usual rule that $(xy)^{-1} = y^{-1}x^{-1}$ holds, since

$$y^{-1}x^{-1}xy = 1 = xyy^{-1}x^{-1}.$$

From (B.3) it follows that

$$x^{-1}xy = 2x^{-1}yx,$$
$$y = 2x^{-1}yx,$$

thus

$$yx^{-1} = 2x^{-1}y.$$

Similarly

$$y^{-1}x = 2xy^{-1}.$$

Thus to multiply x^{-2} by y^3, we have

$$
\begin{aligned}
x^{-2}y^3 &= x^{-1}(x^{-1}y)y^2 \\
&= x^{-1}(\tfrac{1}{2})yx^{-1}y^2 \\
&= (\tfrac{1}{2})(x^{-1}y)(x^{-1}y)y \\
&= (\tfrac{1}{2})(\tfrac{1}{2})yx^{-1}(\tfrac{1}{2})yx^{-1}y \\
&= (\tfrac{1}{8})y(x^{-1}y)(x^{-1}y) \\
&= (\tfrac{1}{32})y(yx^{-1})(yx^{-1}) \\
&= (\tfrac{1}{32})y^2(x^{-1}y)x^{-1} \\
&= (\tfrac{1}{64})y^3x^{-2}.
\end{aligned}
$$

Hilbert defined an element S (as represented in (B.2)) to be greater than 0 if the first coefficient r_0 of T_0 is greater than 0; an element is less than 0 if r_0 is less than 0. Hence the element $-y^3 x + yx^{-1} + 2y^3 x^2 - y^{-4} x^3$ is *less* than 0, since the term $-y^{-4} x^3$ is the leading term when the expression is arranged in terms of ascending exponents of y; the element $-y^2 x^4 + yx^{-1} + 2y^3 x^2 - y^4 x^3$ is *greater* than 0, since its leading term is yx^{-1}. Then an element S_i is greater than S_j if $(S_i - S_j)$ is greater than 0. The lead term in $S_1 - S_2$ given above is $-y^2 x^{-4}$ with coefficient -1; hence S_2 is greater than S_1.

SUGGESTED READING

D. Hilbert, *Foundations of Geometry*, Open Court, LaSalle, IL: 1902. Transl. from the 10th German ed., 1971.

Index

Abbott, J., 185
Abelian group, 213
Absorptive law for lattices, 172
Addition of points on affine lines, 48
Addition properties on affine lines, 55
Additive closure space, 168
Additive identity, 12
Additive inverse, 12
Affine closure, 129
Affine closure space, 164
Affine hull, 129, 165
Affine plane, 10, 18
  $(GF(4))^2$, 80
  $(\mathbb{Z}_5)^2$, 81
  of order 4, 37, 56
  of order 5, 69, 86
  over $\mathbb{D}$, 79
Affine space, 123
  over $\mathbb{Z}_2$, 125
Algebraic lattice, 182
Algebraic view of projective space, 171
Analytic geometry, 6, 11
Antisymmetric relation, 171
Artin, Emil, 204, 210, 211
Associative law, 12
  for lattices, 172
Asymptote, 104
Atom of a lattice, 179
Atomic exchange property, 185
Atomic lattice, 180
Atomistic lattice, 180
Automorphism, inner, 206
Automorphism of a lattice, 176
Automorphism of projective or affine space, 186
Auxiliary point, 55, 64
Axiom, 4
Axiom of Choice, 196
Axioms of Euclid, 4
Axis, 104

Baer, Reinhold, 210, 211
Bennett, M. K., 172, 185
Betten, D., 34, 46
Biatomic lattice, 183
Bijection, 24
Bijection, semilinear, 100
Birkhoff and Mac Lane, 75, 92, 220
Birkhoff, Garrett, 75, 171, 185
Bogart, Kenneth, 46
Bose, Parker and Shrinkhande, 34, 46
Bounded lattice, 175
Bruck and Ryser, 46
Bruck–Ryser Theorem, 39

Cancellation law for addition, 213
Canonical form, 95
Cardinality, 24
Cartesian geometry, 6
Cell, 28
Center of a perspectivity, 197
Central collineation, 191
Centralizer, 166
Chain, 180
Chain with four elements, 177
Characteristic:
  of a division ring, 91
  of a field, 218
Characteristic 0, 219
Characteristic infinity, 219
Closed subset in a closure space, 165
Closure law, 12
Closure operator, 128, 149, 165
Closure space, 128
  additive, 168
  affine or projective, 165
  irreducible, 169
  locally finite, 168
  normalized, 166
  point-closed, 167
  properties of, 167

225